D1045392

FIRST AID MANUAL FOR CHEMICAL ACCIDENTS

Second Edition

Compiled by

Marc J. Lefèvre, M.D.

Hygienist, Toxicologist, Medical Adviser,
Head of the Central Toxicological Service
of the Central Personnel Department of
SOLVAY & Cie., S.A., Brussels

Second English-language edition
revised by

Shirley A. Conibear, M.D., M.P.H.

Associate Professor of Occupational Medicine
University of Illinois College of Medicine,
Chicago

Associate Professor of Environmental
and Occupational Health Sciences
University of Illinois School of Public Health

President, Carnow, Conibear & Associates, Ltd.

VNR VAN NOSTRAND REINHOLD
_____ New York

The recommendations for emergency treatment found in this volume are based on procedures developed by recognized authorities and from the extensive experience of a major chemical producer. Nevertheless, the users of this first aid manual must realize that first aid techniques are constantly being improved and that the general standards for the health and safety of individuals are being raised. It is necessary that the book's users be trained to distinguish correctly the different symptoms of intoxication described in this book, that they be fully qualified to render first aid treatment and that they remain current with the latest developments in first aid treatment. This manual is not to be construed as a substitute for any legal practice. In any case, the author, editor, publisher, or the publisher's agents shall not be liable or responsible in any manner whatsoever for any errors or omissions in this book, or for the use of the contents of this book.

Copyright © 1989 by Van Nostrand Reinhold
Library of Congress Catalog Card Number 88-27996
ISBN 0-442-20490-6

Printed in the United States of America

Van Nostrand Reinhold
115 Fifth Avenue
New York, New York 10003

Van Nostrand Reinhold International Company Limited
11 New Fetter Lane
London EC4P 4EE, England

Van Nostrand Reinhold
102 Dodds Street
South Melbourne 3205, Victoria, Australia

Macmillan of Canada
Division of Canada Publishing Corporation
164 Commander Boulevard
Agincourt, Ontario M1S 3C7, Canada

16 15 14 13 12 11 10 9 8 7 6 5 4 3 2

Libarary of Congress Cataloging-in-Publication Data

Lefèvre, M. J. (Marc J.)
    First aid manual for chemical accidents.

    Translation of: Manuel de premiers soins d'urgence.
    1. First aid in illness and injury. 2. Industrial
accidents. 3. Industrial toxicology. I. Conibear,
Shirley A. II. Title.
RC963.3.L4313 1989   615.9'08   88-27996
ISBN 0-442-20490-6

Contents

Compiler's Preface

This manual of first aid is not the work of only one physician, but of a whole staff.

Without the bibliographical references, books, and data that SOLVAY & Cie has provided to its Department of Toxicology, this manual would have been left as a draft or a project in a drawer.

The first French edition is out of print.

I express all my gratitude for the really valuable assistance to all the people of SOLVAY & Cie, S.A., Brussels of LAPORTE INDUSTRIES, LTD and of THE SOLVAY AMERICAN CORPORATION who collaborated on the issue of this revised English-language edition. The English translation and medical terminology have been reviewed by me.

M. J. LEFÈVRE, M.D.

Editor's Preface

Sudden exposure to industrial chemicals requires prompt, effective action in order to avoid serious injury or even death. This first aid manual provides you with the information needed to make the appropriate response. It is intended for use by people working in factories, laboratories, on farms, or in the transport of industrial chemicals. Law enforcement personnel, firefighters, and others likely to be first on the scene will also find it useful.

This manual was first developed and used by Solvay & Sie, S.A., an international chemical company, to satisfy a practical need to provide specific, easy to locate, and correct advice for workers in various aspects of chemical manufacturing, storage, and transport. The second edition has been extensively expanded to include an Appendix which identifies over 250 pesticides by generic and trade names and an Appendix of the chemicals organized by CAS number. The text itself has been expanded to include more than 200 additional chemicals selected because of their frequent use in the United States.

The book can also be used in non-emergency situations when it is necessary to give advice about emergency actions to others, as in the creation of labels or material safety data sheets. Persons responsible for developing a plan for dealing with emergencies will find it useful, as will those who train others to respond to chemical emergencies. It can be used as a supplement to CPR and first aid training in educational programs for first responders.

The manual is organized to provide a quick index either by common or chemical name, or by CAS number. Once the chemical is identified, the user is directed to a list of expectetd symptoms and then to the proper response depending on the route of exposure. The advice is arranged so that the most urgent problems are taken care of first while stressing to the user the importance of taking appropriate self-protective measures. Often, the victims of industrial accidents tragically multiply as successive rescuers become victims themselves because they have not taken protective measures.

It is hoped that this book will serve as a focal point for organizing a response to exposure emergencies, and will help to minimize or prevent serious injury if an accident does occur. There are many support services available when an accident occurs. The book emphasizes making contact with all parts of the emergency response system including local Emergency Medical Services, and national programs like ChemTrec or Poison Control Centers.

It should be noted that this book deals only with acute exposures. None of the symptoms are considered, nor is the advice appropriate for chronic low level exposures, and the risk of carcinogenesis is not addressed by this book.

Masculine pronouns have been used throughout this book for reasons of conciseness; they are intended to refer to both females and males.

SHIRLEY CONIBEAR, M.D., M.P.H.

INTRODUCTION

I. THE IMPORTANCE OF THE FIRST RESPONDENT'S ROLE

Books and pamphlets on emergency treatment of chemical exposures have usually been written for toxicologists, physicians, or other medical personnel who have years of experience and a thorough knowledge of biochemistry, clinical medicine, toxicology, and diagnostic procedures. Authors also assume that these practitioners have medical equipment and drugs available. Little has been written specifically for use by the first respondent to an accident who is more likely to be the shop foreman or the laboratory employee. This book is written to fill that gap.

The *First Aid Manual for Chemical Accidents* was conceived, planned, and written for persons likely to be first on the accident scene: foremen, production workers, firefighters, chemists, laboratory workers, chemical engineers, agricultural workers, law enforcement personnel, railroad workers, and truck drivers. These are the people, located in the factory, on the production line, on the road, in the laboratory, on farms or in forests, who could suddenly be called upon to render assistance to a fellow worker, a friend, or a stranger who is the victim of chemical poisoning because of a mistake or accident.

The first respondent will have to move the victim to a safe place and revive or sustain vital functions until the arrival of a rescue team possessing the authority, experience and skill to provide definitive emergency medical treatment. It is the first respondent's duty and responsibility to first give the alarm to others in the area, and then in sequence, to protect himself suitably using the appropriate respirator, occlusive clothing, gloves, goggles, etc; to remove the victim from the contaminated area; to notify the Emergency Medical Service (EMS); to take charge of the victim; to support respiration and cardiac function; and to calm and comfort the victim, all with a view to minimizing the consequences of the poisoning. If appropriate procedures including first aid measures are not carried out in a timely fashion, the victim may suffer serious consequences. This book provides you, the first respondent, with specific information about the first aid measures appropriate for over 600 hazardous chemicals in common use.

II. A DESCRIPTION OF THE KNOWLEDGE AND EQUIPMENT PRESUMED TO BE IN THE FIRST RESPONDENT'S POSSESSION

Your proficiency in cardiopulmonary resuscitation (CPR) is presumed. Virtually every community and many companies offer a course in CPR at little or no cost to the individual. Almost everyone is capable of learning this skill and anyone who is serious about wanting to provide help in the case of a chemical exposure should take a CPR course and become proficient. First aid training is desirable and also easy to obtain but is not presumed.

The OSHA Hazard Communication Rule has mandated that in most instances where large volumes of hazardous chemicals are in use, workers be trained in proper protective practices and informed of the nature and hazards of the materials with which they work. Material Safety Data Sheets (MSDS's) and labels are to be supplied and made available in the workplace to aid in identification and proper response in an emergency. However, you, the first respondent, may be from a different department or from outside the company and not be familiar with the materials in use. Ask other workers at the scene for the name of the material and its MSDS or make an attempt to locate and read the label. Trade and common names and chemical synonyms sometimes make identification difficult. For this reason, unique permanent numbers called CAS (Chemical Abstract Service) numbers have been assigned to each individual chemical. These are used internationally and can be used to definitively identify a chemical. A cross reference of chemicals listed in numerical order by their CAS numbers is located in appendix D.

In the case of a fire or explosion, labels and MSDS's may have been destroyed or be unavailable. In that case, you can get clues as to the type of chemical by determining what it was being used for, made with or from, or the final product of the operation. Department of Transportation (DOT) labels may be present on a truck or tank car and give some basic information (see appendix B). Manifests on a truck or train may be available to identify the chemical, the shipper and its destination.

Equipment and supplies to be used in caring for the victim should be selected according to the types of hazards most likely to be encountered in a specific plant, situation, or site. Personal protective equipment including an air-supplied respirator, occlusive clothing, boots, gloves, goggles, etc. must be available to the first respondent and be appropriate for the hazardous products being handled. Other equipment available for emergency care should include a spoon; one or two 8 ounce plastic drinking glasses; a kidney basin; a clean bucket; two pillows; two clean cotton sheets; two blankets, preferably of wool; a stretcher, cot, or bed; access to a sink with hot and cold running water; ice; a cold or, preferably, luke-warm shower; hand soap; towels; a pair of strong blunt scissors; a pocket mask or bag-valve mask for use in giving artificial respiration. For special cases, an oxygen cylinder with control valve and mask may be included.

A few over-the-counter pharmaceutical products should be available such as 70% rubbing alcohol, ethyl alcohol, cotton wool, dry gauze dressing, adhesive tape, lime water, calcined magnesia, activated charcoal, syrup of ipecac and sodium sulfate. Foodstuffs such as milk or fresh eggs are often used. Powdered milk or eggs can be kept well for a long time in hermetically sealed cans. Otherwise, first aid never requires medicinal preparations.

III. ORGANIZATION OF THE MANUAL

Chemical Index of Products

This alphabetical list starting on page 11 gives the chemical and some trade names and other synonyms of almost 600 substances frequently encountered in the chemical industry and in agriculture. In order to avoid making the main index unmanageable, neither a full list of all synonyms nor any pharmaceutical substances proper have been included. This list contains some names of classes or types of chemicals (e.g. alkanes) which do not have a single CAS number. Appendix D contains the entire main list of individual chemicals organized by CAS number. In the case of pesticides, the main alphabetic list is supplemented by Appendix C which contains a list of pesticides in alphabetic order by commercial and common name. This list is cross referenced back to its category or class of pesticide in the main alphabetic list.

The numbers immediately following the chemical names in the main index refer you to the **Signs and Symptoms section (white** pages) and advise first aid measures according to the type(s) of contact with the substances (**yellow for inhalation, green for ingestion, pink for skin contact,** and **blue for eye contact**).

Signs and Symptoms (*white* pages)

In these **white** pages, substances are grouped together in "families" that cause the victim to manifest more or less similar signs and symptoms, with a unique section number assigned to each family. A symptom is something felt by the victim, e.g. shortness of breath. To find out what symptoms the victim has, ask him how and what he feels. A sign is something you can observe about the victim, e.g. rapid, labored breathing. Use your eyes, ears, nose and touch to gain such information about the victim's condition.

At the top of the first page of each chemical family's section is an alphabetic list of names of the chemicals covered in that section. Concise information regarding any special physical or toxic properties follows. Finally, a list of signs and symptoms ordered from mild to increasingly severe appears for each of the various types of contact, with references to the colored pages for first aid. The larger the dose or the longer the exposure, the more likely you are to see signs and symptoms further down the list.

These lists include the more serious signs and symptoms that would be observed or might develop if the victim had a massive exposure and were given no first aid and no subsequent medical treatment. This is why most of the descriptions terminate with the words "coma" and "death."

At this point, one must make a comparison: If a child falls into water or an adult is caught in quicksand, death will inevitably put an end to their vain efforts, their anguish, and their cries for help if no outside aid reaches them. A hand, a cloth, or a branch held out can quickly save their lives or prevent any serious consequences; the accident becomes an incident and eventually an anecdote. But it is necessary that the person seeing the potential tragedy, the first respondent in this case, act quickly and coolly. Reading the descriptions should not, therefore, provoke panic, but instead inspire your determination to act quickly and calmly in order to prevent progression to the most tragic result.

First-Aid Instructions (*yellow, green, pink,* and *blue* pages)

The colored pages list the first aid measures to be given in their prescribed order before the arrival of the EMS rescue team, and depend on the patient's condition, the dose and the effects of the poison.

- The **yellow** pages refer to poisoning by INHALATION.

- The **green** pages refer to poisoning via the digestive tract by INGESTION or SWALLOWING.

- The **pink** pages refer to poisoning by SKIN CONTACT.

- The **blue** pages refer to poisoning by EYE CONTACT.

As a cross check, each of the colored "type-of-contact" sections repeats the list of substances for which the stated first aid measures are appropriate. In order to avoid mistakes, always verify whether the relevant chemical substance is listed at the top of the first-aid page that you are consulting.

Appendix A: General Instructions in Case of Poisoning by Unknown Chemical Products

The information in these pages is intended for cases where the exact chemical name of the injuring substance is doubtful or unknown, for example, a person found unconscious or with evidence of splashing by a mixture of substances. This appendix is also organized by type of contact.

Appendix B: Meaning of DOT Symbols

Vehicles transporting hazardous chemicals must carry a Department of Transportation (DOT) type label which gives some useful information about the chemical being hauled. This appendix tells you how to decipher the DOT symbols found on these labels.

Appendix C: Glossary of Commercial and Common Pesticide Names

This is an alphabetic list of pesticides using commercial or trade names as well as common names. Some pesticides may be listed twice by both commercial and common name. Each is cross referenced back to a pesticide group contained in the main alphabetic listing. This list supplements the main list. Some individual pesticides are also listed individually in the main list. If you don't find the name of the chemical in the main list and you suspect that it is a pesticide, look it up in **APPENDIX C.**

Appendix D: Chemicals in CAS Number Order

This list contains all of the chemicals in the main list that have an individual CAS number.

IV. USING THE MANUAL

To be successful, first aid must be given without delay. To facilitate use of the manual, it is advised that you go through the main list and mark chemicals that are present in your work environment or that you can anticipate may be introduced into the work environment. Fill in the names and phone numbers of emergency response organizations and personnel in the space provided on the last page of this book. Verify these numbers periodically by calling them.

If the chemical substance that has caused the injury is not precisely known, you must make an educated guess as to the identity and nature of the substance. That can be done from the external characteristics (such as smell, appearance, thickness, color, etc.) of the spilled substance, or from the signs and symptoms exhibited by the victim. If you have no idea what the chemical is, consult **APPENDIX A** on page 215.

Generally, however, the substance responsible will be known or discoverable. In that case, follow this procedure:

1. Look up the name of the substance in the main alphabetic chemical index.

The name is followed by several numbers that refer to the sections to be consulted depending on the type of contact. Alternatively, look up the CAS number in **APPENDIX D** if that is the information available to you. Then refer back to the main chemical index. If the material is a pesticide and is not in the main list, check **APPENDIX C** for a category cross reference back to the main product list.

2. Next, turn to the section in the Signs and Symptoms chapter listed in the first column. Verify that the exposure was to a chemical listed in the table at the start of the section. Verify that the victim's symptoms roughly correspond to those described. If they do not, follow the general instructions given in **APPENDIX A** on page 215.

3. If the signs and symptoms do correspond, turn to the section number in the colored pages corresponding to the type of contact that the victim had. Note that skin contact implies inhalation if the material is a liquid that evaporates easily. Suspect skin contact if the material is a gas or vapor, especially if the victim has been sweating or is wet. Ingestion may occur if the victim is exposed to a dust or powder. If he doesn't spit out material coughed up from the lungs, it will end up in the stomach.

4. Quickly read over all the first aid measures to be administered. Verify that you are on the appropriate colored page by again locating the chemical in the table heading the section.

5. Abide by the advice given and follow it in sequence. ALWAYS REMEMBER TO FIRST TAKE THE NECESSARY PRECAUTIONS TO PROTECT YOURSELF FROM EXPOSURE. Remember that sometimes two poisonous effects may occur simultaneously or consecutively. For example:
   a. Splashing with caustic soda lye may affect both the eyes and skin on the arms. In this case, the eyes must take first priority as far as treatment is concerned. Treat them first and then undress the victim under the shower. See the **GOLDEN RULES** on page 8.
   b. A splash of chlorinated solvent soaks a trouser leg. A risk of inhalation can be secondarily associated with the skin contact. The first respondent will take the appropriate action: that is, remove the contaminated clothing to the open air, flush the skin with water and warn those around of the danger.

6. Follow these Golden Rules:

```
┌─────────────────────────────────────────────────────────┐
│                    GOLDEN RULES                         │
│   1. Protect yourself from exposure.                    │
│   2. Terminate the victim's exposure and decontaminate. │
│   3. Always treat the most urgent symptom or sign first:│
│         Cessation of breathing                          │
│         Heart not beating                               │
│         Eye injury                                      │
│         Skin contact                                    │
│         Shock                                           │
│   4. Call for help.                                     │
└─────────────────────────────────────────────────────────┘
```

V. CONCLUSIONS

The *First Aid Manual for Chemical Accidents* is written in clear, concise language using lay terms. It should always be available and ready to be consulted immediately. Familiarize yourself with its contents and organization now, so that you can use it quickly and with assurance. It is intended to contribute to saving lives, if possible; to lessen the suffering of victims of chemical accidents; and to minimize the after effects of exposure.

First respondents will earn the gratitude of the people they help, the esteem of health care professionals, and above all, they will cherish the unforgetable feeling of having unselfishly and competently helped a fellow human being in a moment of tremendous need. We hope that this small book will be carefully read and its message clearly understood.

NOTE

These signs and symptoms are only descriptive of situations involving sudden, intense exposure. Low level, chronic exposure frequently causes a much different set of signs and symptoms and requires different treatment. Absence of the signs and symptoms described here in no way implies that chronic illness due to long-term, low-level exposure is not occurring.

Many of the chemicals listed in this book are known or suspected human carcinogens. Paradoxically, some of these carcinogens have few or mild acute effects. Attention here is always given to decreasing absorption of these chemicals by inducing vomiting or by flushing. A physician should evaluate each of these cases no matter how trivial the exposure seems. The physician will decide on further action in terms of follow-up exams and precautions.

Always refer every victim to a physician for evaluation after administering first aid as described in this book.

CHEMICAL INDEX

		SYMPTOMS	INHALATION	INGESTION	SKIN	EYES	CAS NUMBER
A	Acetaldehyde	1	1	2	1	1	00075-07-7
	Acetic Acid	1	1	1	1	1	00064-19-7
	Acetic Anhydride	1	1	1	1	1	00108-24-7
	Acetone	8	3	12	6	5	00067-64-1
	Acetone Cyanohydrin	25	8	11	8	5	00075-86-5
	Acetonitrile	25	8	11	8	5	00075-05-8
	Acetylene	6	3	0	5	4	00074-86-2
	Acrolein	1	1	2	1	1	00107-02-8
	Acrylamide	17	4	2	6	5	00079-06-1
	Acrylonitrile	25	8	11	8	5	00107-13-1
	Adiponitrile	25	8	11	8	5	00111-69-3
	Aldicarb	23	3	2	11	9	00116-06-3
	Aldrin	31	5	5	2	5	00309-00-2
	Aliphatic Alcohols— Amyl	8	3	12	6	5	00071-41-0
	Aliphatic Alcohols— Butyl	8	3	12	6	5	00071-36-3
	Aliphatic Amines	19	1	3	4	2	00095-38-5
	Alkali Dichromates	3	7	2	2	6	
	Alkali Meta-Borates	3	7	2	2	6	
	Alkanes (gasses, C_1 to C_4)	6	3	0	5	4	
	Alkanes (liquids/solids)	29	2	10	6	6	08002-74-2
	Allyl Alcohol	17	4	2	6	5	00107-18-6
	Allyl Chloride	17	4	2	6	5	00107-05-1
	Allyl Glycidyl Ether	17	4	2	6	5	00106-92-3
	Allyl Propyl Disulfide	17	4	2	6	5	02179-59-1
	Aluminum (dust)	14	6	10	7	7	07429-90-5
	Aluminum Alkyls	28	6	0	10	8	
	Aluminum Chloride	3	7	2	2	6	07446-70-0
	Aluminum Hydrate	14	6	10	7	7	21645-51-2
	Aluminum Hydroxide	14	6	10	7	7	21645-51-2
	Aluminum Oxide	14	6	10	7	7	01344-28-1
	Aluminum Trichloride	3	7	2	2	6	07446-70-0
	Aminopyridine	31	5	5	2	5	00504-29-0
	Ammonia	19	1	3	4	2	07664-41-7
	Ammonium Carbonate	20	2	3	9	3	00506-87-6
	Ammonium Chlorate	12	2	8	2	6	12125-02-9

(continued)

	SYMPTOMS	INHALATION	INGESTION	SKIN	EYES	CAS NUMBER
Ammonium Hydroxide	19	1	3	4	2	01336-21-6
Ammonium Perchlorate	12	2	8	2	6	07790-98-9
Ammonium Sulfide	4	1	3	1	1	12124-99-1
Amyl Acetate	8	3	12	6	5	00628-63-7
Aniline	11	4	7	6	3	00062-53-3
Anisidines (ortho)	11	4	7	6	3	00092-15-9
Anisidines (para)	11	4	7	6	3	00104-94-9
Antimony and Compounds	21	1	2	1	1	07440-36-0
Arsenic	21	1	1	2	6	07440-38-2
Arsenic Trichloride	21	1	1	1	1	07784-34-1
Arsenicals	21	1	1	1	1	
Arsine	21	1	0	1	1	07784-42-1
Asbestos	14	6	10	7	7	01332-21-4
Asphalt Fumes	29	2	0	6	6	
B Barium (soluble salts)	10	2	6	2	6	07440-39-3
Barium Acetate	10	2	6	2	6	00543-80-6
Barium Carbonate	10	2	6	2	6	00513-77-9
Barium Chloride	10	2	6	2	6	10361-37-2
Barium Fluoride	10	2	1	3	1	07787-32-8
Barium Hydroxide	10	2	6	2	6	17194-00-2
Barium Nitrate	10	2	6	2	6	10022-31-8
Barium Oxide	10	2	6	2	6	01304-28-5
Barium Sulfide	10	2	6	2	6	21109-95-5
Benzene	7	3	4	6	5	00071-43-2
Benzidine	11	4	7	6	3	00092-87-5
Benzyl Chloride	1	1	2	1	1	00100-44-7
Bis (Chloromethyl) Ether	17	4	2	6	5	00542-88-1
Bitter Almond Oil (Amygdalin)	25	8	11	8	5	29883-15-6
Boric Acid	3	7	2	2	6	10043-35-3
Boron Trifluoride	1	1	0	1	1	77637-07-2
Bromine	1	1	1	1	1	07726-95-6
Bromoform	9	3	5	6	5	00075-25-2
Butadiene	6	3	0	5	4	00106-99-0
Butane	6	3	0	5	4	00106-97-8
Butanol	8	3	12	6	5	00075-65-0

	SYMPTOMS	INHALATION	INGESTION	SKIN	EYES	CAS NUMBER
Butyl Acetate	8	3	12	6	5	00123-86-4
Butyl Glycidyl Ether (n-)	17	4	2	6	5	02426-08-6
Butylamine	19	1	3	4	2	00109-73-9
Butyltoluene	9	3	5	6	5	00098-51-1
Butyraldehyde	1	1	1	1	1	00123-72-8
C Cadmium (dust and fumes) (metal)	3	7	2	2	6	07440-43-9
Calcium Carbide	5	7	3	4	2	00075-20-7
Calcium Carbonate	14	6	10	7	7	00471-34-1
Calcium Chloride	13	6	8	2	6	10043-52-4
Calcium Dichromate	3	7	2	2	6	
Calcium Hydroxide	20	2	3	9	3	01305-62-0
Calcium Hypochlorite	3	7	1	2	6	07778-54-3
Calcium Oxide	5	7	3	4	2	01305-78-8
Camphor	31	5	5	2	5	00076-22-2
Caprolactam	3	7	2	2	6	00105-60-2
Carbamates	23	3	2	11	9	14484-64-1
Carbon	14	6	10	7	7	07440-44-0
Carbon Black	14	6	10	7	7	01333-86-4
Carbon Dioxide	6	3	0	5	4	00124-38-9
Carbon Dioxide Snow	6	3	0	5	4	
Carbon Disulfide	1	1	2	1	1	00075-15-0
Carbon Monoxide	6	3	0	0	0	00630-08-0
Carbon Tetrachloride	9	3	5	6	5	00056-23-5
Cement	5	7	2	4	2	
Cherry Laurel Water	25	0	11	8	5	
Chlordane	31	5	5	2	5	00057-74-9
Chlorinated Lime	3	7	2	2	6	01332-17-8
Chlorine	1	1	1	1	1	07782-50-5
Chlorine Dioxide	1	1	1	1	1	10049-04-4
Chlorine Trifluoride	2	1	1	3	1	07790-91-2
Chloro-1-Nitropropane (1-)	17	4	2	6	5	00600-25-9
Chloroacetaldehyde	1	1	1	1	1	00107-20-0
Chloroacetic Acid	1	1	1	1	1	00079-11-8
Chloroacetophenone (2-)	17	4	2	6	5	00532-27-4
Chlorobenzene	9	3	5	6	5	00108-90-7

(continued)

	SYMPTOMS	INHALATION	INGESTION	SKIN	EYES	CAS NUMBER
Chlorobenzylidene Malonitrile	17	4	2	6	5	02698-41-1
Chlorobromomethane	9	3	5	6	5	00074-97-5
Chlorodifluoroethane	6	3	0	5	4	00075-68-3
Chlorodifluoromethane	6	3	0	5	4	00075-45-6
Chloroethane	9	3	5	5	4	00075-00-3
Chlorofluoroethane	6	3	0	5	4	00075-68-3
Chlorofluoromethane	6	3	0	5	4	00593-70-4
Chloroform	9	3	5	6	5	00067-66-3
Chloromethane	6	3	0	5	4	00074-87-3
Chloronaphthalenes	29	2	10	6	6	00091-58-7
Chloropentafluoroethane	9	3	0	6	5	00076-15-3
Chlorophenoxy Compounds	26	4	7	6	3	
Chloropicrin	17	4	2	6	5	00076-06-2
Chloropropane	9	3	5	6	5	00075-29-6
Chloropropene	9	3	5	6	5	00557-98-2
Chlorotrifluoroethylene	6	3	0	5	4	00079-38-9
Chlorotrifluoromethane	6	3	0	5	4	00075-72-9
Chlorthion	23	3	2	11	9	00500-28-7
Chromic Acid	3	7	1	2	6	07738-94-5
Chromium Chloride	3	7	2	2	6	10025-73-7
Copper Chloride	3	7	2	2	6	01344-67-8
Copper Sulfate	3	7	2	2	6	07758-98-7
Creosote	1	1	2	1	1	08001-58-9
Creosols	1	1	2	1	1	01319-77-3
Crotonaldehyde	1	1	2	1	1	00123-73-9
Cumene	7	3	4	6	5	00098-82-8
Cyanogen Bromide	25	8	11	8	5	00506-68-3
Cyanogen Chloride	25	8	11	8	5	00506-77-4
Cyanogen Iodine	25	8	11	8	5	00506-78-5
Cyclohexane	7	3	4	6	5	00110-82-7
Cyclohexanol	17	4	2	6	5	00108-93-0
Cyclohexanone	17	4	2	6	5	00108-94-1
D DDVP	23	3	2	11	9	00062-73-7
DNBP	26	4	7	6	3	00088-85-7
DNOC	26	4	7	6	3	00534-52-1

	SYMPTOMS	INHALATION	INGESTION	SKIN	EYES	CAS NUMBER
Decaborane	31	5	5	2	5	17702-41-9
Decane	7	3	4	6	5	00124-18-5
Decanol	8	3	12	6	5	00112-30-1
Demeton	23	3	2	11	9	08065-48-3
Diacetone Alcohol	8	3	12	6	5	00123-42-2
Diazinon	23	3	2	11	9	00333-41-5
Diazomethane	17	4	2	6	5	00334-88-3
Diborane	17	4	0	6	5	23273-02-1
Dibutyl Phthalate	17	4	2	6	5	00084-74-2
Dibutylamine	19	1	3	4	2	00111-92-2
Dibutyllead	18	5	9	6	6	02587-84-0
Dibutyltin	22	5	9	2	6	
Dichloro-5,5-Dimethylhydantoin	1	1	2	1	1	00118-52-5
Dichlorobenzene	9	3	5	6	5	00095-50-1
Dichlorodifluoromethane	6	3	0	5	4	00075-71-8
Dichloroethane	9	3	5	5	4	00075-34-3
Dichloroethylene	9	3	5	6	5	00540-59-0
Dichlorofluoromethane	6	3	0	5	4	00075-43-4
Dichloropropane	9	3	5	6	5	00078-87-5
Dichlorotetrafluoroethane	9	3	5	6	5	00076-14-2
Diepoxybutane	17	4	2	6	5	00564-00-1
Diethylaluminum Chloride	28	6	0	10	8	00096-10-6
Diethylaluminum Hydride	28	6	0	10	8	
Diethylamine	19	1	3	4	2	00109-89-7
Diethylaminoethanol	17	4	2	6	5	00100-37-8
Diethylene Glycol	8	3	12	6	5	00111-46-6
Diethyllead	18	5	9	6	6	15773-47-4
Diethylmercury	15	5	9	2	6	00627-44-1
Diethyltin	22	5	9	2	6	
Difluoroethanes	6	3	0	5	4	00075-37-6
Difluoroethylene	6	3	0	5	4	00075-38-7
Diglycidyl Ether	17	4	2	6	5	02238-07-5
Dihexyltin	22	5	9	2	6	
Diiododiethyltin	22	5	9	2	6	
Diisobutylcarbinol	8	3	12	6	5	00108-82-7

(continued)

	SYMPTOMS	INHALATION	INGESTION	SKIN	EYES	CAS NUMBER
Dimethyl Sulfate	1	1	1	1	1	00077-78-1
Dimethylamine	19	1	3	4	2	00124-40-3
Dimethylaniline	11	4	7	6	3	00121-69-7
Dimethylhydrazine (1,1-)	31	5	5	2	5	00057-14-7
Dimethylmercury	15	5	9	2	6	00593-74-8
Dimethyltin	22	5	9	2	6	
Dinitrobenzene	11	4	7	6	3	25154-54-5
Dinitrocresols	26	4	7	6	3	01335-85-9
Dinitrophenols	26	4	7	6	3	00329-71-5
Dinitrotoluene	11	4	7	6	3	00121-14-2
Dioctyltin	22	5	9	2	6	
Dioxane	8	3	12	6	5	00123-91-1
Diphenyl	29	2	10	6	6	00092-52-4
Diphenylamine	29	2	10	6	6	00122-39-4
Dipropylamine	19	1	3	4	2	00142-84-7
Dipterex	23	3	2	11	9	00052-68-6
Dipyridyl Chloride	30	4	2	6	5	
Dipyridyl Dimethyl Sulfate	30	4	2	6	5	
Diquat	30	4	2	6	5	00085-00-7
Disodium Phosphate	27	6	0	7	6	07558-79-4
Dithiocarbamates	13	6	8	2	6	
E EPN	23	3	2	11	9	02104-64-5
Epichlorohydrin	17	4	2	6	5	00106-89-8
Ethane	6	3	0	5	4	00074-84-0
Ethanolamine	19	1	3	4	2	00141-43-5
Ethyl Acetate	8	3	12	6	5	00141-78-6
Ethyl Acrylate	17	4	2	6	5	00140-88-5
Ethyl Alcohol	8	3	12	6	5	00064-17-5
Ethyl Chloroformate	1	1	2	1	1	00541-41-3
Ethyl Ether	6	3	4	5	4	00060-29-7
Ethyl Fluoride	6	3	0	5	4	00353-36-6
Ethyl Nitrate	26	4	7	6	3	00625-58-1
Ethylamine	19	1	3	4	2	00075-04-7
Ethylbenzene	8	3	12	6	5	00100-41-4
Ethylene	6	3	0	5	4	00074-85-1
Ethylene Chlorohydrin	8	3	12	6	5	00107-07-3

	SYMPTOMS	INHALATION	INGESTION	SKIN	EYES	CAS NUMBER
Ethylene Dichloride	9	3	5	5	4	00107-06-2
Ethylene Glycol	8	3	12	6	5	00107-21-1
Ethylene Glycol Dinitrate	26	4	7	6	3	00628-96-6
Ethylene Glycol Monomethyl Ether	8	3	12	6	5	00109-86-4
Ethylene Oxide	17	4	2	6	5	00075-21-8
Ethyleneimine	17	4	2	6	5	00151-56-4
Ethylhexyl Acetate	8	3	12	6	5	00103-09-3
Ethylmercuric Chloride	15	5	9	2	6	00107-27-7
Ethylmercuric Hydroxide	15	5	9	2	6	
Ethylmercury	15	5	9	2	6	
F Ferricyanides	25	8	11	8	5	
Ferrocyanides	25	8	11	8	5	
Fibrous Glass	14	6	10	7	7	14808-60-7
Fluorine	2	1	1	3	1	07782-41-4
Fluoromethane	6	3	0	5	4	00593-53-3
Fluosilicic Acid	2	1	1	3	1	01309-45-1
Formaldehyde	1	1	2	1	1	00050-00-0
Formic Acid	1	1	1	1	1	00064-18-6
Freon 11	6	3	0	5	4	00075-69-4
Freon 12	6	3	0	5	4	00075-71-8
Freon 13	6	3	0	5	4	00075-72-9
Freon 14	6	3	0	5	4	00075-73-0
Freon 21	6	3	0	5	4	00075-43-4
Freon 22	6	3	0	5	4	00075-45-6
Freon 112	9	3	5	6	5	00076-12-0
Freon 113	9	3	5	6	5	00076-13-1
Freon 114	9	3	5	6	5	00076-14-2
Freon 115	9	3	5	6	5	00076-15-3
Freon 116	6	3	0	5	4	00076-16-4
Freon 142b	6	3	0	5	4	00075-68-3
Freon 143	6	3	0	5	4	
Freon 151a	6	3	0	5	4	
Freon 152a	6	3	0	5	4	
Furfural	17	4	2	6	5	00098-01-1
Furfural Alcohol	8	3	12	6	5	00098-00-0

(continued)

		SYMPTOMS	INHALATION	INGESTION	SKIN	EYES	CAS NUMBER
G	Gasoline	7	3	4	6	5	08006-61-9
	Glutaraldehyde	17	4	2	6	5	00111-30-8
	Glycerin	8	3	12	6	5	00056-81-5
	Glycidol	17	4	2	6	5	00556-52-5
	Glycidyl Acrylate	17	4	2	6	5	00106-90-1
	Gramoxone	30	4	2	6	5	04685-14-7
H	Halothane	9	3	5	6	5	00074-96-4
	Heptane	7	3	4	6	5	00142-82-5
	Heptanol	8	3	12	6	5	00543-49-7
	Hexachlorobenzene	29	2	10	6	6	00118-74-1
	Hexachloroethane	9	3	5	6	5	00067-72-1
	Hexafluoroethane	6	3	0	5	4	00076-16-4
	Hexane	7	3	4	6	5	00110-54-3
	Hexanol	8	3	12	6	5	00111-27-3
	Hydrazine	31	4	5	6	5	00302-01-2
	Hydriodic Acid	1	1	1	1	1	10034-85-2
	Hydrochloric Acid	1	1	1	1	1	07647-01-0
	Hydrocyanic Acid	25	8	11	8	5	00074-90-8
	Hydrofluoric Acid	2	1	1	3	1	07664-39-3
	Hydrogen Bromide	1	1	1	1	1	10035-10-6
	Hydrogen Chloride	1	1	1	1	1	07647-01-0
	Hydrogen Peroxide	1	1	1	1	1	07722-84-1
	Hydrogen Selenide	4	1	0	1	1	07783-07-5
	Hydrogen Sulfide	4	1	0	1	1	07783-06-4
	Hydroquinone	11	4	7	6	3	00123-31-9
I	Iodine	1	1	2	1	1	07553-56-2
	Iron Chloride	3	7	2	2	6	07758-94-3
	Isobutyl Acetate	8	3	12	6	5	00110-19-0
	Isobutyraldehyde	1	1	1	1	1	00078-84-2
	Isobutyronitrile	25	8	11	8	5	00078-82-0
	Isopestox	23	3	2	11	9	00371-86-8
	Isopropyl Acetate	8	3	12	6	5	00108-21-4
	Isopropyl Alcohol	8	3	12	6	5	00067-63-0
	Isopropylamine	19	1	3	4	2	00075-31-0
K	Kaolin	14	6	10	7	7	01332-58-7
	Ketene	17	4	0	6	5	00463-51-4

		SYMPTOMS	INHALATION	INGESTION	SKIN	EYES	CAS NUMBER
L	Lead (dust and fumes)	24	6	2	2	6	07439-92-1
	Lead Acetate	24	6	2	2	6	15347-57-6
	Lead Antimonate	24	6	2	2	6	
	Lead Arsenate	24	6	2	2	6	07784-40-9
	Lead Carbonate	24	6	2	2	6	00598-63-0
	Lead Chromate	24	6	2	2	6	18454-12-1
	Lead Chromate (yellow)	24	6	2	2	6	07758-97-6
	Lead Dioxide	24	6	2	2	6	01309-60-0
	Lead Nitrate	24	6	2	2	6	10099-74-8
	Lead Oleate	18	5	9	6	6	01120-46-3
	Lead Oxide (PbO)	24	6	2	2	6	01309-60-0
	Lead Oxide (red)	24	6	2	2	6	01314-41-6
	Lead Oxychloride	24	6	2	2	6	
	Lead Phenate	18	5	9	6	6	
	Lead Phthalate	18	5	9	6	6	
	Lead Stearate	18	5	9	6	6	07428-48-0
	Lead Subacetate	24	6	2	2	6	01335-32-6
	Lead Sulfide	24	6	2	2	6	01314-87-0
	Leptophos	23	3	2	11	9	21609-90-5
	Lime	5	7	3	4	2	01305-78-8
	Lindane	31	5	5	2	5	00058-89-9
	Liquified Petroleum (LP) Gas	6	3	4	5	4	68476-85-7
	Lithium Carbonate	20	2	3	9	3	00554-13-2
	Lithium Hydride	20	2	3	9	3	07580-67-8
M	Magnesium Chloride	27	6	0	7	6	07786-30-3
	Magnesium Sulfate	27	6	0	7	6	07487-88-9
	Malathion	23	3	2	11	9	00121-75-5
	Maleic Anhydride	1	1	2	1	1	00108-31-6
	Malononitrile	25	8	11	8	5	00109-77-3
	Mercuric Chloride	16	5	2	2	6	07487-94-7
	Mercuric Iodine (red)	16	5	2	2	6	07774-29-0
	Mercurous Chloride	16	5	2	2	6	07546-30-7
	Mercurous Iodide	16	5	2	2	6	07783-30-4
	Mercury (metal)	16	5	2	2	6	07439-97-6

(continued)

	SYMPTOMS	INHALATION	INGESTION	SKIN	EYES	CAS NUMBER
Mercury (organic compounds)	15	5	9	2	6	
Mercury (soluble salts)	16	5	2	2	6	07439-97-6
Mercury Acetate	16	5	2	2	6	00631-60-7
Mercury Fulminate	15	5	9	2	6	00628-86-4
Mercury Nitrate	16	5	2	2	6	10045-94-0
Mercury Oxycyanide	16	5	2	2	6	01335-31-5
Methacrylonitrile	25	8	11	8	5	00126-98-7
Methane	6	3	0	5	4	00074-82-8
Methyl Acetate	8	3	12	6	5	00079-20-9
Methyl Acrylate	17	4	2	6	5	00096-33-3
Methyl Alcohol	8	3	12	6	5	00067-56-1
Methyl Bromide	17	4	0	6	5	00074-83-9
Methyl n-Butyl Ketone	8	3	12	6	5	00591-78-6
Methyl Chloride	6	3	0	5	4	00074-87-3
Methyl Chloroformate	1	1	2	1	1	00079-22-1
Methyl Ethyl Ketone	8	3	12	6	5	00078-93-3
Methyl Isoamyl Ketone	8	3	12	6	5	00110-12-3
Methyl Isobutyl Ketone	8	3	12	6	5	00108-10-1
Methyl Isocyanate	17	4	2	6	5	00624-83-9
Methyl Isopropyl Ketone	8	3	12	6	5	00563-80-4
Methyl Mercaptan	4	1	0	1	1	00074-93-1
Methyl Methacrylate Monomer	17	4	2	6	5	00080-62-6
Methyl Nitrate	26	4	7	6	3	00598-58-3
Methyl Parathion	23	3	2	11	9	00298-00-0
Methylamine	19	1	3	4	2	00074-89-5
Methylchloroform	9	3	5	6	5	00071-55-6
Methylene Chloride	9	3	5	5	4	00075-09-2
Methylene Fluoride	6	3	0	5	4	
Methylenebis (Phenyl Isocyanate)	17	4	2	6	5	00101-68-8
Methylmercury	15	5	9	2	6	22967-92-6
Methylmercury Borate	15	5	9	2	6	
Methylmercury Hydroxide	15	5	9	2	6	01184-57-2
Methylmercury Iodide	15	5	9	2	6	
Methylmercury Nitrate	15	5	9	2	6	

		SYMPTOMS	INHALATION	INGESTION	SKIN	EYES	CAS NUMBER
	Methylmercury Phosphate	15	5	9	2	6	32787-44-3
	Milk of Lime	20	0	3	9	2	
	Monomethylhydrazine	11	4	7	6	3	00060-34-4
N	Naphthalene	11	4	7	6	3	00091-20-3
	Naphthylamines	11	4	7	6	3	00091-59-8
	Naptha	7	3	4	6	5	08030-30-6
	Nickel (fumes and dust)	17	4	2	6	5	07440-02-0
	Nickel Carbonyl	17	4	2	6	5	13463-39-3
	Nitric Acid	1	1	1	1	1	07697-37-2
	Nitric Oxide	1	1	0	1	1	10102-43-9
	Nitroanilines	11	4	7	6	3	00100-01-6
	Nitrobenzene	11	4	7	6	3	00098-95-3
	Nitrochlorobenzene (p-)	11	4	7	6	3	00100-00-5
	Nitrocresolic Herbicides	26	4	7	6	3	
	Nitroferricyanides (salts)	25	8	11	8	5	
	Nitrogen	6	3	0	5	4	07727-37-9
	Nitrogen Dioxide	1	1	1	1	1	10102-44-0
	Nitrogen Trifluoride	1	1	0	3	1	07783-54-2
	Nitroglycerine	26	4	7	6	3	00055-63-0
	Nitromethane	7	3	4	6	5	00075-52-5
	Nitrophenolic Herbicides	26	4	7	6	3	00554-84-7
	Nitrophenols	26	4	7	6	3	00554-84-7
	Nitrotoluene	11	4	7	6	3	00088-72-2
	Nonane	7	3	4	6	5	00111-84-2
O	OMPA	23	3	2	11	9	00152-16-9
	Octane	7	3	4	6	5	00111-86-4
	Organochlorines	31	5	5	2	5	
	Organophosphate Compounds	23	3	2	11	9	
	Osmic Acid	1	1	1	1	1	20816-12-0
	Oxalic Acid	1	2	1	1	1	00144-62-7
	Oxygen Difluoride	2	1	0	3	1	07783-41-7
	Ozone	1	1	0	1	1	10028-15-6
P	Paraffins	29	2	10	0	6	08002-74-2

(continued)

	SYMPTOMS	INHALATION	INGESTION	SKIN	EYES	CAS NUMBER
Paraoxon	23	3	2	11	9	00311-45-5
Paraquat	30	4	5	6	5	01910-42-5
Parathion	23	3	2	11	9	00056-38-2
Pentaborane	31	5	5	2	5	19624-22-7
Pentachloroethane	9	3	5	6	5	00076-01-7
Pentachlorophenate	26	4	7	6	3	00087-86-5
Pentachlorophenol	26	4	7	6	3	00087-86-5
Pentane	7	3	4	6	5	00109-66-0
Pentanol	8	3	12	6	5	06032-29-7
Peracetic Acid	1	1	1	1	1	00079-21-0
Perborates	3	7	2	2	6	
Perchloric Acid	1	1	1	1	1	07601-90-3
Perchloromethyl Mercaptan	17	4	2	6	5	00594-42-3
Perchloryl Fluoride	1	1	0	3	1	07616-94-6
Petroleum Ethers	7	3	4	6	5	08030-30-6
Phenol	1	1	1	1	1	00108-95-2
Phenylenediamine (p-)	1	1	2	1	1	00106-50-6
Phenylhydrazine	11	4	7	6	3	00100-63-0
Phenylhydroxylamine	11	4	7	6	3	00100-65-2
Phenylmercuric Acetate	15	5	9	2	6	00062-38-4
Phenylmercury	15	5	9	2	6	
Phenylmercury Oleate	15	5	9	2	6	
Phenylnaphthylamine	11	4	7	6	3	00135-88-6
Phorate	23	3	2	11	9	00298-02-2
Phosdrin	23	3	2	11	9	07786-34-7
Phosgene	1	1	0	1	1	00075-44-5
Phosphine	21	1	0	1	1	07803-51-2
Phosphoric Acid	1	1	1	1	1	07664-38-2
Phosphoric Ester	23	3	2	11	9	
Phosphorus	21	1	1	1	1	07723-14-0
Phosphorus Chlorides	21	1	1	1	1	10025-87-3
Phosphorus Pentachloride	21	1	1	1	1	10026-13-8
Phosphorus Pentasulfide	4	1	0	1	1	01314-80-3
Phosphorus Trichloride	21	1	1	1	1	07719-12-2
Phthalic Anhydride	3	7	2	2	6	00085-44-9
Picric Acid	17	4	2	6	5	00088-89-1
Platinum and Compounds	3	7	2	2	6	07440-06-4

	SYMPTOMS	INHALATION	INGESTION	SKIN	EYES	CAS NUMBER
Polybrominated Biphenyls (PBBs)	29	2	10	6	6	36355-01-8
Polychlorinated Biphenyls (PCBs)	29	2	10	6	6	01336-36-3
Polyvinyl Chloride	14	6	10	7	7	00075-01-4
Potassium	5	7	3	4	2	07440-09-7
Potassium Carbonate	20	2	3	9	3	00584-08-7
Potassium Chlorate	12	2	8	2	6	03811-04-9
Potassium Chloride	13	6	8	2	6	07477-40-7
Potassium Chlorite	1	1	1	1	1	
Potassium Chromate	3	7	2	2	6	07789-00-6
Potassium Cyanide	25	8	11	8	5	00151-50-8
Potassium Dichromate	3	7	2	2	6	07778-50-9
Potassium Fluoride	2	1	1	3	1	07789-23-3
Potassium Fluosilicate	2	7	1	3	1	16871-90-2
Potassium Hydroxide	5	7	3	4	2	01310-58-3
Potassium Oxide	5	7	3	4	2	
Potassium Perchlorate	12	2	8	2	6	07778-74-7
Propane	6	3	0	5	4	00074-98-6
Propionaldehyde	1	1	1	1	1	00123-38-6
Propyl Acetate	8	3	12	6	5	00109-60-4
Propyl Alcohol	8	3	12	6	5	00071-23-8
Propyl Nitrate	26	4	7	6	3	00627-13-4
Propylamine	19	1	3	4	2	00107-10-8
Propylene	6	3	0	5	4	00115-07-1
Propylene Glycol	8	3	12	6	5	00057-55-6
Propylene Glycol Monomethyl Ether	17	4	2	6	5	00107-98-2
Propylene Oxide	17	4	2	6	5	00075-56-9
Pyrethrins	17	4	2	6	5	
Q Quaternary Ammonium Compounds	30	4	2	6	5	
Quinone	1	1	2	1	1	00106-51-4
R Resorcinol	1	1	1	1	1	00108-46-3
Ronnel	23	3	2	11	9	00299-84-3

(continued)

		SYMPTOMS	INHALATION	INGESTION	SKIN	EYES	CAS NUMBER
S	Selenium Hexafluoride	4	1	0	1	1	07783-79-1
	Silane	1	1	0	1	1	07803-62-5
	Silica	14	6	10	7	7	07631-86-9
	Sodium	5	7	3	4	2	07440-23-5
	Sodium Bicarbonate	13	6	8	2	6	00144-55-8
	Sodium Borate	3	7	2	2	6	01303-96-4
	Sodium Carbonate	20	2	3	9	3	00497-19-8
	Sodium Chlorate	12	2	8	2	6	07775-09-9
	Sodium Chloride	13	6	8	2	6	07647-14-5
	Sodium Chlorite	1	1	1	1	1	07758-19-2
	Sodium Chromate	3	7	2	2	6	10034-82-9
	Sodium Cyanide	25	8	11	8	5	13998-03-3
	Sodium Dichromate	3	7	2	2	6	07789-12-0
	Sodium Fluoride	2	1	1	3	1	07681-49-4
	Sodium Fluosilicate	2	7	1	3	1	16893-85-9
	Sodium Hydroxide	5	7	3	4	2	01310-73-2
	Sodium Hypochlorite	3	7	2	2	6	07681-52-9
	Sodium Oxide	5	7	3	4	2	01313-59-3
	Sodium Perchlorate	12	2	8	2	6	07601-89-0
	Sodium Peroxide	5	7	3	4	2	01313-60-6
	Sodium Silicate	20	2	3	9	3	06834-92-0
	Sodium Sulfate	27	6	0	7	6	07767-82-6
	Sodium Thiocyanate	27	6	0	8	5	00540-72-7
	Sodium Thiosulfate	27	6	0	7	6	07772-98-7
	Stibine	21	1	0	1	1	07803-52-3
	Stoddard Solvent	7	3	4	6	5	08052-41-3
	Styrene	17	4	2	6	5	00100-42-5
	Sulfotepp	23	3	2	11	9	03689-24-5
	Sulfur Dioxide	1	1	1	1	1	07446-09-5
	Sulfur Trioxide	1	1	1	1	1	07446-11-9
	Sulfuric Acid	1	1	1	1	1	07664-93-9
	Sulfurous Acid	1	1	1	1	1	07782-99-2
T	TEPP	23	3	2	11	9	00107-49-3
	Talc	14	6	10	7	7	14807-96-6
	Tellurium Hexafluoride	1	1	0	1	1	07783-80-4
	Tetrabutyltin	22	5	9	2	6	01461-25-2
	Tetrachlorodifluoroethane	9	3	5	6	5	00076-11-9

	SYMPTOMS	INHALATION	INGESTION	SKIN	EYES	CAS NUMBER
Tetrachloroethane	9	3	5	6	5	00079-34-5
Tetrachloroethylene	9	3	5	6	5	00127-18-4
Tetraethyllead	18	5	9	6	6	00078-00-2
Tetraethyltin	22	5	9	2	6	00597-64-8
Tetrafluoroethylene	6	3	0	5	4	00116-14-3
Tetrafluoromethane	6	3	0	5	4	00075-73-0
Tetraisoalkyltin	22	5	9	2	6	
Tetramethyl Succinonitrile	25	8	11	8	5	03333-52-6
Tetramethyllead	18	5	9	6	6	00075-74-1
Tetranitromethane	11	4	7	6	3	00509-14-8
Tetrapentyltin	22	5	9	2	6	
Tetrapropyltin	22	5	9	2	6	02176-98-9
Tetryl	17	4	2	6	5	00479-45-8
Thiocarbamates	13	6	8	2	6	
Titanium (dust and fumes)	14	6	10	7	7	07440-32-6
Titanium Chlorides	3	7	2	2	6	07550-45-0
Titanium Dioxide	14	6	10	7	7	13463-67-7
Tolidine (o-)	11	4	7	6	3	00119-93-7
Toluene	7	3	4	6	5	00108-88-3
Toluene 2,4-di-Isocyanate	17	4	2	6	5	00584-84-9
Toluene 2,6-di-Isocyanate	17	4	2	6	5	00091-08-7
Toluidine	11	4	7	6	3	00095-53-4
Tributyl Phosphate	1	1	2	1	1	00126-73-8
Tributyllead	18	5	9	6	6	
Tributyltin	22	5	9	2	6	
Trichloroacetic Acid	1	1	1	1	1	00076-03-9
Trichloroethane	9	3	5	6	5	00079-00-5
Trichloroethylene	9	3	5	6	5	00079-01-6
Trichlorofluoromethane	6	3	0	5	4	00075-69-4
Trichlorofluoroethane	9	3	5	6	5	00076-13-1
Triethylaluminum	28	6	0	10	8	00097-93-8
Triethylamine	19	1	3	4	2	00121-44-8
Triethylene Glycol	8	3	12	6	5	00112-27-6
Triethyllead	18	5	9	6	6	00562-95-8
Trifluoroethane	6	3	4	5	4	27987-06-0

(*continued*)

		SYMPTOMS	INHALATION	INGESTION	SKIN	EYES	CAS NUMBER
	Trifluoromethane	6	3	0	5	4	00075-46-7
	Triisobutylaluminum	28	6	0	10	8	00100-99-2
	Trimellitic Anhydride	3	7	2	2	6	00552-30-7
	Trimethylaluminum	28	6	0	10	8	00075-24-1
	Trimethylamine	19	1	3	4	2	00075-50-3
	Trimethyllead	18	5	9	6	6	07442-13-9
	Trimethyltin	22	5	9	2	6	
	Trinitrobenzene	11	4	7	6	3	00099-35-4
	Trinitrotoluene	11	4	7	6	3	00118-96-7
	Triphenyltin	22	5	9	2	6	00892-20-6
	Tripropyltin	22	5	9	2	6	00761-44-4
	Trisodium Phosphate	20	2	3	9	3	07601-54-9
	Trithion	23	3	2	11	9	00786-19-6
	Tungsten Carbide	14	6	10	7	7	12070-12-1
	Turpentine	7	3	4	6	5	08006-64-2
U	Uranium Compounds	3	7	2	2	6	07440-61-1
V	Vanadium and Compounds	3	7	2	2	6	07440-62-2
	Vinyl Acetate	8	3	12	6	5	00108-05-4
	Vinyl Chloride	9	3	0	6	5	00075-01-4
	Vinyl Fluoride	6	3	0	5	4	00075-02-5
	Vinylidene Chloride	9	3	5	5	4	00075-35-4
	Vinylidene Fluoride	6	3	0	5	4	00075-38-7
X	Xylene	7	3	4	6	5	01330-20-7
	Xylidine	11	4	7	6	3	00095-68-1
Y	Yttrium and Compounds	14	6	10	7	7	10361-92-9
Z	Zinc Chloride	3	7	2	2	6	07646-85-7

DESCRIPTION OF
SYMPTOMS OF
POISONING

Acetaldehyde
Acetic Acid
Acetic Anhydride
Acrolein
Benzyl Chloride
Boron Trifluoride
Bromine
Butyraldehyde
Carbon Disulfide
Chlorine
Chlorine Dioxide
Chloroacetaldehyde
Chloroacetic Acid
Creosote
Cresols
Crotonaldehyde
Dichloro-5,5-Dimethylhydantoin
Dimethyl Sulfate
Ethyl Chloroformate
Formaldehyde
Formic Acid
Hydriodic Acid
Hydrochloric Acid
Hydrogen Bromide
Hydrogen Chloride
Hydrogen Peroxide
Iodine
Isobutyraldehyde
Maleic Anhydride

Methyl Chloroformate
Nitric Acid
Nitric Oxide
Nitrogen Dioxide
Nitrogen Trifluoride
Osmic Acid
Oxalic Acid
Ozone
Peracetic Acid
Perchloric Acid
Perchloryl Fluoride
Phenol
Phenylenediamine (p-)
Phosgene
Phosphoric Acid
Potassium Chlorite
Propionaldehyde
Quinone
Resorcinol
Silane
Sodium Chlorite
Sulfur Dioxide
Sulfur Trioxide
Sulfuric Acid
Sulfurous Acid
Tellurium Hexafluoride
Tributyl Phosphate
Trichloroacetic Acid

The above caustic (corrosive) substances all give off irritating vapors or fumes as soon as they come into contact with mucous membranes or moist skin. The more soluble in water they are, the quicker the symptoms develop. Chlorites give off CHLORINE DIOXIDE as soon as they come into contact with dilute acid. The concentration and composition of these products will determine the seriousness of injury.

I. INHALATION

A. Signs and symptoms of acute poisoning

1. Irritation of mucous membranes (nose, mouth, eyes, throat)

2. Watering of eyes, nasal discharge, sneezing, coughing (head cold-like)

3. Oppressive feeling in the chest or chest pain

4. Difficulty in breathing, wheezing

5. Rapid breathing

6. Fits of coughing

7. Headache

8. BLUISH face and lips

9. Salivation

10. Giddiness

11. Nausea

12. Muscular weakness

13. Ulceration of mucous membranes (nose)

14. Fluid in the lungs (pulmonary edema)

15. Secondary chemical pneumonia

16. Death

B. *First aid:* **see** *yellow* **pages section** *1*, **or see** *yellow* **pages, section** *2* **(OXALIC ACID only)**

II. INGESTION

A. Signs and symptoms of acute poisoning

1. Irritation and a burning sensation of lips, mouth, and throat

2. Pain in swallowing

3. Abundant salivation or drooling

4. Ulceration of the mucous membranes of the mouth and color change of the tongue (GRAY with HYDROCHLORIC ACID, YELLOW with NITRIC ACID, WHITE to BLACK with SULFURIC ACID, WHITE with HYDROGEN PEROXIDE, OXALIC ACID, and PHENOL)

5. Intense thirst

6. Swelling in the throat

7. Burning sensation in the chest (esophagus), back of throat, and stomach

8. Painful abdominal cramps

9. Nausea and vomiting, occasionally of coffee-grounds-like material (digestive bleeding)

10. Difficulty in breathing

11. Risk of perforation of the stomach causing a hard, rigid abdomen

12. State of shock
    - weak and rapid pulse
    - cold sweat—pale complexion
    - lightheadedness
    - cold hands and feet

13. Convulsions

14. Coma

15. Death

B. *First aid:* **see** *green* **pages, section** *1,* **or** *green* **pages, section** *2,* **per Chemical Index**

III. SKIN CONTACT

PHENOL, which penetrates healthy skin, may cause generalized poisoning, resulting in death.

Vapors are generally less irritating to the skin than solutions, but the concentration and the length of contact determine the gravity of the symptoms.

In an accident where volatile products come in contact with the skin, poisoning by inhalation will frequently also occur.

A. Immediate signs and symptoms

1. Smarting or tingling

2. Burning sensation

3. Redness and swelling

4. Burns that may be very painful and become WHITE (in the case of

HYDROCHLORIC ACID, HYDROGEN PEROXIDE, and PHENOL) or YELLOW (in the case of NITRIC ACID).

5. Painful blisters

6. Profound damage to tissues (painless in the case of PHENOL)

7. Shock can occur as a result of pain
   - weak and rapid pulse
   - cold sweat—pale complexion
   - lightheadedness
   - cold hands and feet

8. Coma

9. Death

10. Lingering death (PHENOL only)

B. *First aid:* see *pink* pages, section *1*, or *pink* pages, section *3*, per Chemical Index

IV. SPLASHING OF LIQUID IN OR CONTACT OF VAPOR WITH EYES

In an accident caused by volatile products, poisoning by inhalation will frequently also occur.

A. **Immediate symptoms**

1. Stinging or burning sensation

2. Watering of eyes

3. Redness and swelling of eyelids

4. Intense pain in eyelids and eyes, with ulceration of the tissues

5. Yellow eyes (NITRIC ACID only)

6. Opaqueness of the cornea causing blurred vision

7. Loss of sight

B. *First aid:* see *blue* pages, section *1*

Chlorine Trifluoride

Fluorine

Fluosilicic Acid

Hydrofluoric Acid

Oxygen Difluoride

Potassium Fluoride

Potassium Fluosilicate

Sodium Fluoride

Sodium Fluosilicate

HYDROFLUORIC ACID is a very strong acid with a highly caustic and corrosive effect on organic tissue. The resulting burns often do not become painful until several hours later. Fluosilicates are less irritating but are toxic to liver and kidneys.

I. INHALATION

A. Signs and symptoms of acute poisoning

1. Irritation of mucous membranes (nose, eyes, mouth, throat)

2. Watering of eyes

3. Salivation

4. Fits of coughing

5. Ulceration of mucous membranes (nose and throat)

6. Pain in throat

7. Chest heaviness or pain

8. Difficulty in breathing

9. Bronchitis

10. Headache

11. Fatigue

12. Giddiness

13. Nausea

14. Stomach pains

15. Pale or BLUISH face

16. State of shock

- weak and rapid pulse
- cold sweat—pale complexion
- lightheadedness
- cold hands and feet

17. Fluid in the lungs (acute pulmonary edema)

18. Coma

19. Death

B. *First aid:* see *yellow* pages, section *1*, or see *yellow* pages, section *7* (FLUOSILICATES only)

II. INGESTION

A. Signs and symptoms of acute poisoning

1. Irritation and painful burning sensation of lips, mouth, and throat

2. Ulceration of mucous membranes

3. Pain in swallowing

4. Swelling in the throat

5. Intense thirst

6. Violent burning sensation in the chest (esophagus) and stomach

7. Painful stomach cramps with distension of stomach

8. Difficulty in breathing

9. Muscular fatigue and general weakness

10. Coughing

11. Fluid in the lungs (acute pulmonary edema)

12. Perforation of stomach causing a hard, rigid abdomen

13. Nausea and vomiting, occasionally of coffee-grounds-like material (digestive hemorrhage)

14. Pale or BLUISH face

15. State of shock

- weak and rapid pulse
- cold sweat—pale complexion
- lightheadedness
- cold hands and feet

16. Convulsions

17. Coma

18. Death

B. *First aid:* **see** *green* **pages, section** *1*

III. SKIN CONTACT

A. Immediate signs and symptoms

1. Burning sensation and redness and swelling

2. Irreparable damage to skin, which gradually turns white and becomes very painful

3. Blisters

4. Profound damage deep in the tissues

5. Shock may occur as a result of pain
   - weak and rapid pulse
   - cold sweat—pale complexion
   - lightheadedness
   - cold hands and feet

B. *First aid:* **see** *pink* **pages, section** *3*

IV. SPLASHING OF LIQUID IN OR CONTACT OF VAPOR WITH EYES

A. Immediate signs and symptoms

1. Intense stinging and burning sensation

2. Watering of eyes

3. Redness and swelling of eyelids

4. Burning sensation in eyelids and eyes, with ulceration of the tissues

5. Irreparable damage to cornea with blurred vision

6. Loss of vision

B. *First aid:* **see** *blue* **pages, section** *1*

Alkali Dichromates	Perborates
Alkali Meta-Borates	Phthalic Anhydride
Aluminum Chloride	Platinum and Compounds
Aluminum Trichloride	Potassium Chromate
Boric Acid	Potassium Dichromate
Cadmium (dust and fumes) (metal)	Sodium Borate
Calcium Dichromate	Sodium Chromate
Calcium Hypochlorite	Sodium Dichromate
Caprolactam	Sodium Hypochlorite
Chlorinated Lime	Titanium Chlorides
Chromic Acid	Trimellitic Anhydride
Chromium Chloride	Uranium and Compounds
Copper Chloride	Vanadium and Compounds
Copper Sulfate	Zinc Chloride
Iron Chloride	

In water, these substances are all irritating to mucous membranes of the digestive system and eyes, in addition to the specific toxicity of their cations.

ALUMINUM TRICHLORIDE and TITANIUM TETRACHLORIDE hydrolyze rapidly and in the presence of water, produce HYDROGEN CHLORIDE (see **white** pages, section **1**).

I. INHALATION OF DUST

A. Signs and symptoms of very acute poisoning

1. Irritation of nose and eyes

2. Fits of coughing, sometimes violent

3. Difficulty in breathing

4. BLUISH face and lips

5. Risk of fluid in the lungs (pulmonary edema)

B. Signs and symptoms of acute poisoning

1. Irritation of nose and eyes

2. Tingling and burning sensation in lungs

3. Sneezing

C. *First aid:* see *yellow* pages, section *7*

II. INGESTION

A. Signs and symptoms of acute poisoning

1. Irritation of mouth and throat

2. Salivation

3. Burning sensation in stomach

4. Stomach cramps

5. Nausea and vomiting, occasionally of coffee-grounds-like material (digestive hemorrhage)

6. General weakness, dizziness

7. Diarrhea, possibly blood-stained

8. State of shock
   - weak and rapid pulse
   - cold sweat—pale complexion
   - lightheadedness
   - cold hands and feet

9. Possible convulsions

10. Risk of paralysis (MAGNESIUM CHLORIDE only)

11. Coma

12. Death

B. *First aid:* see *green* pages, section *2*, or *green* pages, section *1*, per Chemical Index

III. CONTACT WITH MOIST SKIN

A. Immediate signs and symptoms

1. Itching

2. Irritation and burning sensation

3. Redness and swelling

4. Ulceration and possibly profound necrosis (destruction of tissue)

B. *First aid:* see *pink* pages, section *2*

IV. SPLASHING IN EYES

A. Immediate signs and symptoms

1. Stinging and burning sensation

2. Watering of eyes

3. Redness and swelling of eyelids

4. Risk of serious injury

B. *First aid:* see *blue* pages, section *6*

Ammonium Sulfide*

Hydrogen Selenide

Hydrogen Sulfide

Methyl Mercaptan

Phosphorus Pentasulfide

Selenium Hexafluoride

These are colorless gases with a strong smell of garlic or rotten eggs, highly toxic and irritating to mucous membranes of the respiratory system and eyes. The nose is very sensitive and can smell small amounts. However, the presence of a large quantity of the gas results in inability to smell it.

* AMMONIUM SULFIDE and its solutions decompose readily into Hydrogen Sulfide.

I. INHALATION

A. Signs and symptoms of very acute poisoning

1. Sudden loss of consciousness

2. Breathing stops abruptly

3. Death follows quickly

B. Signs and symptoms of acute poisoning

1. Irritation of nose, throat, and eyes

2. Sneezing

3. Headache

4. Excitability

5. Dizziness, staggering

6. Nausea and vomiting

7. Breathing difficulties

8. Pale complexion

9. Dry cough

10. Cold sweat

11. Diarrhea

12. Muscular weakness

13. Drowsiness

14. Risk of fluid in the lungs (pulmonary edema)

15. Death

B. *First aid:* see *yellow* **pages, section** *1*

II. INGESTION

No risk, except for AMMONIUM SULFIDE

First aid: see **green** pages, section **3**

III. SKIN CONTACT

The vapors cause no adverse symptoms on the skin, even in toxic concentrations. The solutions may be irritating.

A. Immediate signs and symptoms

1. Slight irritation

2. Painful inflammation

3. Possible dark discoloration

B. *First aid:* see *pink* **pages, section** *1*

IV. CONTACT OF VAPOR WITH EYES

A. Immediate signs and symptoms

1. Irritation

2. Watering of eyes

3. Redness and swelling of eyelids

4. Risk of serious injury

B. *First aid:* see *blue* **pages, section** *1*

Calcium Carbide*   Potassium Oxide
Calcium Oxide   Sodium
Cement   Sodium Hydroxide
Lime   Sodium Oxide
Potassium   Sodium Peroxide
Potassium Hydroxide

These substances are particularly caustic (corrosive) in contact with mucous membranes and moist skin. They all give off heat when dissolved in water. This increases their corrosiveness since they are capable of dissolving living tissue.

*When CALCIUM CARBIDE comes into contact with moist skin, it decomposes into CALCIUM OXIDE, HYDROXIDE, and ACETYLENE, while PHOSPHINE and ARSINE are formed from impurities (see **white** pages, section **21**).

I. INHALATION OF DUST

A. Signs and symptoms of very acute poisoning

1. Irritation of nose, eyes, and throat

2. Burning sensation in nose and throat

3. Difficulty in breathing

4. Fits of coughing

5. Risk of fluid in the lungs (acute pulmonary edema)

B. Signs and symptoms of acute poisoning

1. Irritation of nose and eyes

2. Sneezing

3. Tingling sensation in nose and throat

4. Cough

C. *First aid:* see *yellow* pages, section 7

II. INGESTION

A. Signs and symptoms of acute poisoning

1. Immediate, intense burning sensation in mouth, throat, and stomach

2. Immediate effect on cheeks and mouth membranes, which become white

3. Intense pain in swallowing

4. Swelling of the throat

5. Extreme salivation and drooling, which increases the pain

6 Vomiting; vomit is coffee-grounds-like and contains pieces of mucous membrane of the stomach (digestive hemorrhage)

7. Stomach cramps

8. Rapid breathing

9. State of shock
   - weak and rapid pulse
   - cold sweat—pale complexion
   - lightheadedness
   - cold hands and feet

10. Diarrhea, possibly blood-stained

11. Risk of perforation of stomach causing a hard, rigid abdomen

12. Loss of consciousness

13. Death

B. *First aid:* see *green* pages, section *3,* or *green* pages, section *2,* per Chemical Index

III. SKIN CONTACT

A. Immediate signs and symptoms

1. Itching and tingling sensation

2. Painful sensation

3. Painful ulceration with a slippery, soapy feeling

4. Profound, irreparable damage to tissues

5. State of shock
   - weak and rapid pulse
   - cold sweat—pale complexion
   - lightheadedness
   - cold hands and feet

B. *First aid:* see *pink* pages, section *4*

IV. SPLASHING IN EYES

A. Immediate or delayed signs and symptoms

1. Highly painful, instantaneous irritation of eyes and eyelids

2. Intense watering of eyes

3. Victim keeps eyelids tightly closed

4. Burns and irreparable damage of mucous membranes

5. Ulceration of eyes

6. Perforation of eyes and eyelids

7. Loss of eyes or eyesight

NOTE: If only one eye is splashed and first aid is not given immediately, the other eye may be lost as well, even if the substance has not come into contact with it!

B. *First aid:* see *blue* pages, section *2*

Acetylene*	Ethyl Fluoride
Alkanes (gasses, C_1 to C_4)	Ethylene
Butadiene	Fluoromethane
Butane	Freon 11, 12, 13, 14, 21, 22, 116,
Carbon Dioxide	142b, 143, 151a, 152a
Carbon Dioxide Snow	Hexafluoroethane
Carbon Monoxide	Liquefied Petroleum Gas
Chlorodifluoroethane	Methane
Chlorodifluoromethane	Methyl Chloride
Chlorofluoroethane	Methylene Fluoride
Chlorofluoromethane	Nitrogen
Chloromethane	Propane
Chlorotrifluoroethylene	Propylene
Chlorotrifluoromethane	Tetrafluoroethylene
Dichlorodifluoromethane	Tetrafluoromethane
Dichlorofluoromethane	Trichlorofluoromethane
Difluoroethanes	Trifluoroethane
Difluoroethylene	Trifluoromethane
Ethane	Vinyl Fluoride
Ethyl Ether	Vinylidene Fluoride

Most of these substances are gaseous or evaporate readily at room temperature, and are practically nonirritating and, with the exception of CARBON MONOXIDE, and METHYL CHLORIDE, nontoxic as such. However, in high concentrations they all cause asphyxia by displacing oxygen. METHYL CHLORIDE causes lesions in the cornea due to inflammation.

*In the case of technical grade ACETYLENE, highly toxic impurities such as ARSINE and PHOSPHINE (see **white** pages, section **21**) should be taken into account.

I. INHALATION

A. Signs and symptoms following severe exposure

ASPHYXIA

1. Need for fresh air or "air hunger"

2. Rapid, occasionally irregular, breathing

3. Headache

4. Fatigue

5. Mental confusion

6. Nausea and vomiting

7. Giddiness and poor judgement

8. Exhaustion

9. Loss of consciousness

10. Convulsions

11. Death

B. *First aid:* **see** *yellow* **pages, secton** *3*

II. INGESTION

Practically no risk for the gases. For the liquids:

A. Signs and symptoms of severe poisoning

1. Breath has ethereal sweetish smell

2. Mental confusion

3. Drowsiness

4. Headache

5. Loss of consciousness

6. Death

B. *First aid:* **see** *green* **pages, section** *4*

III. SKIN CONTACT

Splashes of the liquid substances or sprays of the gases that have a low boiling point may "freeze" the skin. Inhalation of the vapors is likely.

A. Immediate signs and symptoms

1. Feeling of intense cold

2. Insensitivity to pain

3. Skin looks white and feels hard and cold, rapidly becoming painful (freezing)

B. *First aid:* see *pink* pages, section 5

IV. SPLASHING OF LIQUID IN EYES

The vapors have no serious irritating effect on the eyes.

When liquids that have a low boiling point are splashed in the eyes, they may cause severe "freezing."

A. Immediate or delayed signs and symptoms

1. Stinging pain

2. Watering of eyes

3. Inflammation of eye lids

4. Cloudiness or opaqueness of eyes

B. *First aid:* see *blue* pages, section 4

Benzene	Nonane
Cumene	Octane
Cyclohexane	Pentane
Decane	Petroleum Ethers
Gasoline	Stoddard Solvent
Heptane	Toluene
Hexane	Turpentine
Naptha	Xylene
Nitromethane	

The vapors of these substances are occasionally irritating to mucous membranes of the respiratory system and eyes and, in high concentrations, they are narcotic.

I. INHALATION

A. Signs and symptoms of acute poisoning

a. by vapors

1. Rapid breathing

2. Excitability, with drunken behavior and mental confusion

3. Staggering

4. Headache

5. Fatigue

6. Nausea with vomiting

7. Dizziness

8. Drowsiness

9. Narcosis (stupor and unresponsiveness)

10. Loss of consciousness

11. Convulsions

12. Coma and death

b. by mist or fine droplets

1. Coughing

2. Difficulty in breathing

3. BLUISH face and lips

4. Nausea and vomiting

5. Fatigue

6. Fever and cough indicative of chemical bronchitis or pneumonia

7. Fluid in the lungs (acute pulmonary edema)

B. *First aid:* see *yellow* pages, section *3*

II. INGESTION

A. Signs and symptoms of acute poisoning

1. Slight gastro-intestinal irritation

2. Dizziness

3. Fatigue

4. Loss of consciousness

5. Coma and death

   If the victim recovers, there is risk of chemical pneumonia as evidenced by:

6. Coughing

7. Fever

B. *First aid:* see *green* pages, section *4*

III. SKIN CONTACT

These gaseous substances have no effect on the skin. In liquid form, they remove oils from the skin, which eventually leads to the following:

A. Delayed signs and symptoms

1. Dryness

2. Cracking and irritation of the skin

B. *First aid:* see *pink* pages, section *6*

IV. SPLASHING OF LIQUID IN OR CONTACT OF VAPOR WITH EYES

The vapors of these substances are only slightly irritating to the eyes. Liquids are strongly irritating to the eyes and to mucous membranes.

A. Immediate signs and symptoms

1. Stinging sensation

2. Watering of eyes

3. Inflammation of the eyelids

B. *First aid:* see *blue* pages, section *5*

Acetone	Glycerin
Aliphatic Alcohols—Amyl	Heptanol
Aliphatic Alcohols—Butyl	Hexanol
Amyl Acetate	Isobutyl Acetate
Butanol	Isopropyl Acetate
Butyl Acetate	Isopropyl Alcohol
Decanol	Methyl Acetate
Diacetone Alcohol	Methyl Alcohol
Diethylene Glycol	Methyl n-Butyl Ketone
Diisobutylcarbinol	Methyl Ethyl Ketone
Dioxane	Methyl Isoamyl Ketone
Ethyl Acetate	Methyl Isobutyl Ketone
Ethyl Alcohol	Methyl Isopropyl Ketone
Ethylbenzene	Pentanol
Ethylene Chlorohydrin	Propyl Acetate
Ethylene Glycol	Propyl Alcohol
Ethylene Glycol Monomethyl Ether	Propylene Glycol
Ethylhexyl Acetate	Triethylene Glycol
Furfuryl Alcohol	Vinyl Acetate

These liquids are more or less volatile and, in high concentrations, cause depression of the brain and nervous system and damage to the liver.

GLYCERIN is only slightly volatile and reputed to be nontoxic.

I. INHALATION

A. Signs and symptoms of acute poisoning

1. Slight irritation of nose and eyes

2. Head feels hot and face is flushed

3. Excitability and talkativeness

4. Drunken behavior

5. Staggering and lack of coordination

6. Headache

7. Mental confusion and visual disturbance

8. Tiredness

9. Nausea and vomiting

10. Paleness of complexion

11. Dizziness

12. Eyes are sensitive to and painful in direct light (METHYL ALCOHOL only)

13. Drowsiness

14. Stupor

15. Loss of consciousness

16. Coma and death

NOTE: METHYL ALCOHOL can seriously impair vision and may cause blindness.

B. *First aid:* see *yellow* pages, section *3*

II. INGESTION

A. Signs and symptoms of acute poisoning

1. Gastro-intestinal irritation

2. Followed by symptoms described in I. INHALATION

NOTE: METHYL ALCOHOL can seriously impair vision and may cause blindness. Glycol derivatives are toxic to the kidneys.

B. *First aid:* see *green* pages, section *12*

III. SKIN CONTACT

A. Immediate signs and symptoms

● Highly volatile products (METHYL ALCOHOL and ACETONE) produce a feeling of cold.

- Alcohols, acetates, and ACETONE remove oils from the skin, which becomes dry and eventually develops cracks or dermatitis.

- Glycol derivatives have little effect on healthy skin.

NOTE: METHYL ALCOHOL, which can be absorbed by the skin, causes
  - headache
  - fatigue
  - reduction of visual acuity

B. *First aid:* see *pink* pages, section *6*

IV. SPLASHING IN EYES

A. Immediate signs and symptoms

1. Vapors are slightly uncomfortable; splashes are irritating

2. Irritation with painful burning or stinging sensation

3. Watering of eyes

4. Inflammation of the eyelids

5. Eyes are sensitive to and painful in the light

B. *First aid:* see *blue* pages, section *5*

Bromoform	Ethylene Dichloride
Butyltoluene	Freon 112, 113, 114, 115
Carbon Tetrachloride	Halothane
Chlorobenzene	Hexachloroethane
Chlorobromomethane	Methylchloroform
Chloroethane	Methylene Chloride
Chloroform	Pentachloroethane
Chloropentafluoroethane	Tetrachlorodifluoroethane
Chloropropane	Tetrachloroethane
Chloropropene	Tetrachloroethylene
Dichlorobenzene	Trichloroethane
Dichloroethane	Trichloroethylene
Dichloroethylene	Trichlorotrifluoroethane
Dichloropropane	Vinyl Chloride
Dichlorotetrafluoroethane	Vinylidene Chloride

The chlorinated and chlorofluorinated solvents are colorless, volatile gases or liquids that smell of chloroform. They are, in varying degrees, toxic to the nervous system, liver, and kidney. They remove oils from skin, which eventually becomes dry and cracked. METHYLENE CHLORIDE is converted to carbon monoxide in the body.

The fluorinated solvents have little irritating or toxic effect on the nervous system, but can cause irregular heart beat.

I. INHALATION

A. Signs and symptoms of acute poisoning

1. Irritation of eyes, nose, and throat

2. Overexcitement

3. Headache

4. Drunkenness (staggering, loss of equilibrium)

5. Irregular heart beat resulting in sudden death

6. Loss of consciousness

7. Narcosis (deep unconsciousness)

8. State of shock

- weak and rapid pulse
- cold sweat—pale complexion
- lightheadedness
- cold hands and feet

9. Coma

10. Death as a result of cardiac or respiratory failure

NOTE: There is risk of fluid in the lungs (acute pulmonary edema) if the victim recovers.

B. Signs and symptoms of relatively acute poisoning

1. Headache

2. Fatigue, weariness

3. Nausea, vomiting

4. Dizziness

5. Irregular heart beat resulting in sudden death

6. Stupor

7. Drowsiness

8. Disturbed vision

9. Coughing

10. Narcosis (deep unconsciousness)

C. *First aid:* see *yellow* pages, section *3*

II. INGESTION

(Except for chlorinated solvents with low boiling points.)

A. Symptoms of acute poisoning

CHARACTERISTIC SYMPTOM: breath smells of chloroform

1. Irritation of lips and mouth

2. Gastro-intestinal irritation

3. Irregular heart beat resulting in sudden death

4. Nausea, vomiting

5. Diarrhea, possibly blood-stained

6. Drowsiness

7. Loss of consciousness

8. Narcosis (deep unconsciousness)

9. State of shock
   - weak and rapid pulse
   - cold sweat—pale complexion
   - lightheadedness
   - cold hands and feet

10. Risk of fluid in the lungs (acute pulmonary edema)

B. *First aid:* see *green* **pages, section 5**

III. SKIN CONTACT

Chlorinated solvents absorbed through healthy skin may cause symptoms described under I. INHALATION, B.

A. Immediate and delayed signs and symptoms

1. Dry skin

2. Freezing of skin

3. Redness and swelling

4. Blisters that later become painful

B. *First aid:* see *pink* **pages, section 6, or** *pink* **pages, section 5, per Chemical Index**

IV. SPLASHING OF LIQUID IN OR CONTACT OF VAPOR WITH EYES

A. Immediate symptoms caused by vapors

1. Irritation of eyes

2. Watering of eyes

3. Occasionally inflammation of eyelids

B. Immediate signs and symptoms when liquid is splashed in eyes

1. Burning sensation

2. Watering of eyes

3. Inflammation of eyelids

4. Lesions in cornea due to inflammation (VINYL CHLORIDE)

C. *First aid:* **see** *blue* **pages, section 5 or** *blue* **pages, section 4, per Chemical Index**

Barium (soluble salts) Barium Hydroxide
Barium Acetate Barium Nitrate
Barium Carbonate Barium Oxide
Barium Chloride Barium Sulfide
Barium Fluoride

All the soluble barium compounds have a toxic effect on the nervous system and heart and may lead to death in the course of a few hours. Soluble barium compounds occur as crystals and have a disagreeable, styptic taste.
NOTE: Only BARIUM SULFATE is insoluble and nontoxic.

I. INHALATION

Soluble barium compounds in the form of dust may be absorbed by nasal mucous membranes and cause a moderate degree of poisoning.

A. Signs and symptoms of acute poisoning

1. Irritation of nose and eyes (BARIUM FLUORIDE, BARIUM HYDROX-IDE, and BARIUM OXIDE)

2. Muscular twitching

3. Tendency to fatigue

4. Abdominal cramps

5. Cold sweat

6. Slow heart rate

7. Death from heart stopping

B. *First aid:* **see** *yellow* **pages, section** *2*

II. INGESTION

A. Symptoms of acute poisoning

1. Disagreeable taste

2. Muscular twitching

3. Nausea and vomiting

4. Stomach pains and diarrhea

5. Anxiety

6. Slow heart rate

7. Convulsions

8. BLUISH face and lips

9. State of shock
   - weak and rapid pulse
   - cold sweat—pale complexion
   - lightheadedness
   - cold hands and feet

10. Paralysis of lower limbs, spreading to upper limbs

11. Difficulty in breathing

12. Death by respiratory failure or respiratory paralysis or heart stoppage

B. *First aid:* **see** *green* **pages, section** *6,* **or** *green* **pages, section** *1*
(BARIUM FLUORIDE only)

III. SKIN CONTACT

BARIUM FLUORIDE, BARIUM OXIDE, and soluble salts are irritating to the skin.

A. Symptoms and signs

1. Irritation of skin and mucous membranes

2. Prolonged contact of dusts with moist skin causes
   - ulcerations
   - necrosis (destruction of tissue) (BARIUM FLUORIDE only)

B. *First aid:* see *pink* pages, section *2*, or *pink* pages, section *3* (BARIUM FLUORIDE only)

IV. SPLASHING IN EYES

A. Immediate signs and symptoms

1. Mechanical and chemical irritation

2. Watering of eyes

3. Inflammation of eyelids (BARIUM FLUORIDE, BARIUM OXIDE, and BARIUM HYDROXIDE)

4. Burning sensation

B. *First aid:* see *blue* pages, section *6*, or *blue* pages, section *1* (BARIUM FLUORIDE only)

Aniline	Nitrobenzene
Anisidines (ortho)	Nitrochlorobenzene (p-)
Anisidines (para)	Nitrotoluene
Benzidine	Phenylhydrazine
Dimethylaniline	Phenylhydroxylamine
Dinitrobenzene	Phenylnaphthylamine
Dinitrotoluene	Tetranitromethane
Hydroquinone	Tolidine (o-)
Monomethylhydrazine	Toluidine
Naphthalene	Trinitrobenzene
Naphthylamines	Trinitrotoluene
Nitroanilines	Xylidine

These substances are toxic to the blood and prevent hemoglobin from carrying oxygen.

I. INHALATION

A. Symptoms of acute poisoning

1. Face, lips, and hands are DEEP BLUE

2. Headache

3. Rapid and difficult breathing

4. Dizziness

5. Mental confusion

6. General weakness

7. Convulsions

8. Coma

9. Death

B. *First aid:* see *yellow* pages, section *4*

II. INGESTION

A. Signs and symptoms of acute poisoning

See I. INHALATION

1. Irritation of mouth and stomach

2. Stomach cramps and diarrhea

B. *First aid:* see *green* **pages, section 7**

III. SKIN CONTACT

Those substances that are fat-soluble can be absorbed by healthy skin and cause the same symptoms as described under I. INHALATION.

A. Additional immediate signs and symptoms

1. Irritation

2. Small vesicles or blisters

3. Redness and swelling

4. Ulceration and necrosis (destruction of tissue)

B. *First aid:* see *pink* **pages, section 6**

IV. SPLASHING IN EYES

A. Immediate symptoms

1. Irritation

2. Redness and swelling of eyelids

3. Pain when looking into light

4. Severe damage to eyes

B. *First aid:* see *blue* **pages, section 3**

Ammonium Chlorate	Potassium Perchlorate
Ammonium Perchlorate	Sodium Chlorate
Potassium Chlorate	Sodium Perchlorate

These substances come in the form of white, highly reactive crystals.
The perchlorates, and the chlorates even more so, are toxic to the blood.
They prevent hemoglobin from carrying oxygen. After 1 to 3 days, they also
cause serious lesions in the liver and kidneys that may result in death.

I. INHALATION

There is little risk of poisoning because the serious risk of fire requires
strict safety measures dealing with the formation of dust.

A. Signs and symptoms of acute poisoning

1. Irritation of nose and eyes

2. Sneezing

3. Small ulcerations in the nose

B. *First aid:* see *yellow* pages, section *2*

II. INGESTION

A. Signs and symptoms of acute poisoning

1. Stomach pains

2. Nausea and vomiting

3. Diarrhea

4. BLUISH face and hands

5. Rapid breathing

6. Dizziness

7. Mental confusion

8. General weakness

9. Coma

10. Death

B. *First aid:* see *green* pages, section *8*

III. SKIN CONTACT

Clothes that have simply been dried must not be returned to the victim. It is essential that they be washed with plenty of water.

NOTE: Clothes that have been impregnated with chlorate or perchlorate solution may, once dry, burst spontaneously into flames; they may also burn as a result of simple friction or contact with a spark or hot cigarette ash.

A. Immediate signs and symptoms

1. Slight irritation

2. Inflammation

3. Ulcerations may follow later

4. Burns

B. *First aid:* see *pink* pages, section *2*

IV. SPLASHING IN EYES

A. Immediate signs and symptoms

1. Mechanical irritation

2. Watering of eyes

3. Inflammation of eyelids

4. Burns

B. *First aid:* see *blue* pages, section *6*

13

Calcium Chloride

Dithiocarbamates

Potassium Chloride

Sodium Bicarbonate

Sodium Chloride

Thiocarbamates

These soluble salts have a very low level of toxicity.

Their concentrated aqueous solutions, however, may be slightly caustic (corrosive) and affect mucous membranes of the digestive system and eyes.

I. INHALATION

Only when large amounts of fine dust are inhaled will the following symptoms occur:

A. Symptoms of acute poisoning

1. Slight irritation of nose

2. Sneezing

B. *First aid:* **see** *yellow* **pages, section** *6*

II. INGESTION

A. Symptoms and signs of acute poisoning

1. Disagreeable taste

2. Nausea and vomiting

3. Gastro-intestinal irritation (CALCIUM CHLORIDE only)

4. Fever

5. Muscle twitching

6. Seizure

7. Rapid respirations

8. Slow heart beat

9. Death

B. *First aid:* **see** *green* **pages, section** *8*

III. SKIN CONTACT

A. Immediate signs and symptoms

1. Irritation

2. Inflammation

3. Small ulcerations

B. *First aid:* see *pink* pages, section *2*

IV. SPLASHING IN EYES

A. Immediate signs and symptoms

1. Mechanical irritation

2. Watering of eyes

3. Inflammation of eye lids

B. *First aid:* see *blue* pages, section *6*

14

Aluminum (dust)	Kaolin
Aluminum Hydrate	Polyvinyl Chloride
Aluminum Hydroxide	Silica
Aluminum Oxide	Talc
Asbestos	Titanium (dust and fumes)
Calcium Carbonate	Titanium Dioxide
Carbon	Tungsten Carbide
Carbon Black	Yttrium and Compounds
Fibrous Glass	

These substances are solid, amorphous, and have few immediate toxic symptoms. POLYVINYL CHLORIDE decomposes under heat, giving off hydrochloric acid fumes (see HYDROCHLORIC ACID, **white** pages, section **1**.)

ASBESTOS and SILICA have been recognized in the United States as having serious injurious effects on the lungs in both long and short ranges, particularly the former. Criteria documents discussing both from many aspects are available from the Superintendent of Documents, U.S. Government Printing Office, Washington, D.C. 20402.

OCCUPATIONAL EXPOSURE TO ASBESTOS—HSM—72—10267, 2nd printing, 1972.

OCCUPATIONAL EXPOSURE TO CRYSTALLINE SILICA—HEW, Publication No. (NIOSH) 75-120, 1974.

OCCUPATIONAL RESPIRATORY DISEASES (edited by J.A. Merchant) DHHS (NIOSH), Publication No. 86-102, 1986.

I. INHALATION

A. Symptoms and signs of heavy inhalation

1. Sneezing

2. Slight irritation of nose

3. Cough

B. *First aid:* see *yellow* pages, section *6*

II. INGESTION

No symptoms

A. *First aid:* **see** *green* **pages, section** *10*

III. SKIN CONTACT

A. Immediate signs and symptoms

1. Itching and redness (FIBROUS GLASS)

B. *First aid:* **see** *pink* **pages, section** *7*

IV. SPLASHING IN EYES

A. Immediate symptoms

1. Mechanical irritation and pain
2. Watering of eyes
3. Inflammation of eyelids

B. *First aid:* **see** *blue* **pages, section** *7*

Diethylmercury
Diemethylmercury
Ethylmercuric Chloride
Ethylmercuric Hydroxide
Ethylmercury
Mercury (organic compounds)
Mercury Fulminate
Methylmercury

Methylmercury Borate
Methylmercury Hydroxide
Methylmercury Iodide
Methylmercury Nitrate
Methylmercury Phosphate
Phenylmercuric Acetate
Phenylmercury
Phenylmercury Oleate

These substances are all toxic on inhalation, ingestion, or contact with skin. They cause lesions in many organs, and particularly in the nervous system.

They are toxic on acute or chronic exposure.

All cases of poisoning by these substances, even if they are not serious, must be reported to the physician, who alone is competent to definitively treat the victim.

I. INHALATION

A. Signs and symptoms of acute poisoning

1. Metallic taste

2. Salivation

3. Headache

4. Dizziness

5. Clumsiness, lack of coordination

6. General weakness

7. Tremor in arms and legs (occasionally)

B. *First aid:* see *yellow* pages, section 5

II. INGESTION

In addition to the symptoms mentioned under I. INHALATION

A. Signs and symptoms of acute poisoning

1. Irritation of mouth and throat
2. Stomach pains
3. Nausea and vomiting
4. Diarrhea

B. *First aid:* see *green* **pages, section** *9*

III. SKIN CONTACT

All organic derivatives of mercury are absorbed by healthy skin and may cause chronic poisoning.

A. Immediate symptoms

1. Irritation
2. Itching
3. Redness
4. Blisters
5. Ulcerations (MERCURY FULMINATE only)

B. *First aid:* see *pink* **pages, section** *2*

IV. SPLASHING IN EYES

A. Immediate signs and symptoms

1. Irritation
2. Watering of eyes
3. Irritation of eyelids
4. Serious lesions may occur in eyes

B. *First aid:* see *blue* **pages, section** *6*

Mercuric Chloride
Mercuric Iodide (red)
Mercurous Chloride
Mercurous Iodide
Mercury (metal)

Mercury (soluble salts)
Mercury Acetate
Mercury Nitrate
Mercury Oxycyanide

Metallic mercury's vapors are sufficiently volatile to cause very severe poisoning. The soluble mercury salts are also all toxic to tissues and, in addition to their severe acute toxicity, cause chronic mercury poisoning on exposure over long periods. Chronic mercury poisoning has highly characteristic symptoms but is not discussed in this manual.

I. INHALATION

A. Signs and symptoms of very acute poisoning (mercury vapors)

1. Metallic taste

2. Rapid and difficult breathing

3. Coughing

4. Bronchitis followed by chemical pneumonia

5. Risk of fluid in the lungs (acute pulmonary edema) for up to 15 days

B. Signs and symptoms of acute poisoning after 2 days

1. Evil-smelling salivation

2. Inflammation of mouth

3. Profuse sweating

4. Headache

5. Nausea, vomiting

6. Stomach cramps

7. Diarrhea

8. Great weakness

9. Tremors

C. *First aid:* see *yellow* pages, section 5

II. INGESTION

A. Immediate signs and symptoms

1. Metallic taste

2. Intense thirst

3. Pain in swallowing

4. Stomach or abdominal pain

5. Nausea and vomiting, vomit may contain blood or greenish substance

6. Diarrhea, occasionally blood-stained or greenish

7. State of shock
   - weak and rapid pulse
   - cold sweat—pale complexion
   - lightheadedness
   - cold hands and feet

8. Tremor in arms and legs

B. *First aid:* see *green* pages, section *2*

III. SKIN CONTACT

The soluble mercury salts are irritating to the skin. They may also penetrate healthy skin.

Metallic mercury is not irritating but may be absorbed through healthy skin.

A. Immediate signs and symptoms

1. Irritation

2. Inflammation

3. Blisters

B. *First aid:* see *pink* pages, section *2*

IV. SPLASHING IN EYES

Metallic mercury is not irritating. Insoluble mercury salts are mechanical irritants while soluble salts are chemical irritants to the mucous membranes.

A. Immediate signs and symptoms

1. Irritation

2. Watering of eyes

3. Inflammation of eyelids

4. Swelling of eyelids

5. Occasionally serious lesions in eyes

B. *First aid:* see *blue* pages, section *6*

Acrylamide	Ethyleneimine
Allyl Alcohol	Furfural
Allyl Chloride	Glutaraldehyde
Allyl Glycidyl Ether	Glycidol
Allyl Propyl Disulfide	Glycidyl Acrylate
Bis(Chloromethyl) Ether	Ketene
Butyl Glycidyl Ether (n-)	Methyl Acrylate
Chloro-1-Nitropropane (1-)	Methyl Bromide
Chloroacetophenone (2-)	Methyl Isocyanate
Chlorobenzylidene Malonitrile	Methyl Methacrylate Monomer
Chloropicrin	Methylenebis(Phenyl Isocyanate)
Cyclohexanol	Nickel (fumes and dust)
Cyclohexanone	Nickel Carbonyl
Diazomethane	Perchloromethyl Mercaptan
Diborane	Picric Acid
Dibutyl Phthalate	Propylene Glycol Monomethyl Ether
Diepoxybutane	Propylene Oxide
Diethylaminoethanol	Pyrethrins
Diglycidyl Ether	Styrene
Epichlorohydrin	Tetryl
Ethyl Acrylate	Toluene 2,4-di-Isocyanate
Ethylene Oxide	Toluene 2,6-di-Isocyanate

These substances are all irritating to the respiratory organs, skin, and mucous membranes of the digestive system and eyes. Some of them have a toxic effect on the liver and kidneys. Some are narcotic in high concentration.

I. INHALATION

A. Signs and symptoms of very acute poisoning

1. Irritation of mucous membranes (nose, eyes, throat)

2. Watering of eyes and nose

3. Chest heaviness or pain

4. Difficulty in breathing

5. Coughing

6. BLUISH face and lips

7. Wheezing

8. Risk of fluid in the lungs (acute pulmonary edema)

B. Signs and symptoms of acute poisoning

1. Irritation of mucous membranes (nose, eyes)

2. Headache

3. Nausea, occasionally vomiting

4. BLUISH face and lips

5. Dizziness

6. Fatigue

7. Diarrhea

C. *First aid:* see *yellow* pages, section *4*

II. INGESTION

Because of their pungent smell, there is little risk of ingestion of these substances.

A. Signs and symptoms of acute poisoning

1. Irritation of lips, mouth, and throat

2. Pain in swallowing

3. Stomach and abdominal pain

4. Nausea and vomiting

5. Diarrhea

6. State of shock
   - weak and rapid pulse
   - cold sweat—pale complexion
   - lightheadedness
   - cold hands and feet

7. Convulsions may occur

8. Risk of fluid in the lungs (acute pulmonary edema)

B. *First aid:* **see** *green* **pages, section** *2*

III. CONTACT WITH MOIST SKIN

These substances all penetrate healthy skin and are highly irritating to moist skin.

A. Immediate signs and symptoms

1. Itching and irritation (except with ETHYLENE OXIDE and PROPYLENE OXIDE)

2. Inflammation

3. Blisters that may be very large, although painless at first

4. Burns and painful ulceration (NICKEL CARBONYL and ALLYL ALCOHOL only)

B. *First aid:* **see** *pink* **pages, section** *6*

IV. SPLASHING IN EYES

A. Immediate signs and symptoms

1. Irritation

2. Watering of eyes

3. Inflammation of eyelids

4. Chemical burns in corneas with pain and difficulty seeing

5. Possibility of serious lesions in eyes

B. *First aid:* **see** *blue* **pages, section** *5*

Dibutyllead	Tetraethyllead*
Diethyllead	Tetramethyllead*
Lead Oleate	Tributyllead*
Lead Phenate	Triethyllead*
Lead Phthalate	Trimethyllead*
Lead Stearate	

These organic lead compounds can all be absorbed by healthy skin. In cases of severe exposure, they are toxic to the nervous system, liver, and kidneys.

They are highly caustic (corrosive), volatile liquids or gases.

I. INHALATION

The risk of severe poisoning depends on the volatility of the compounds and is greatest in those compounds with an asterisk (*).

A. Signs and symptoms of acute poisoning

1. Occasional irritation of nose

2. Trembling

3. Restlessness

4. Delirium

5. Convulsions

B. Signs and symptoms of mild poisoning

1. Restlessness

2. Trembling of hands

3. Headache

4. Pale complexion

5. Occasionally nausea and vomiting

6. Insomnia and nightmares

C. *First aid:* see *yellow* pages, section 5

II. INGESTION

A. Signs and symptoms of acute poisoning

See I. INHALATION

B. *First aid:* see *green* pages, section *9*

III. SKIN CONTACT

A. Immediate signs and symptoms

1. Itching
2. Inflammation
3. Blisters

B. *First aid:* see *pink* pages, section *6*

IV. SPLASHING IN EYES

A. Immediate signs and symptoms

1. Irritation
2. Watering of eyes
3. Inflammation of eyelids

B. *First aid:* see *blue* pages, section *6*

Aliphatic Amines	Ethanolamine
Ammonia	Ethylamine
Ammonium Hydroxide	Isopropylamine
Butylamine	Methylamine
Dibutylamine	Propylamine
Diethylamine	Triethylamine
Dimethylamine	Trimethylamine
Dipropylamine	

Aqueous solutions of these substances are highly irritating to mucous membranes of the respiratory and digestive systems, eyes, and to healthy skin.

I. INHALATION

A. Signs and symptoms of very acute poisoning

1. Irritation of nose, throat, and eyes

2. Burning sensation

3. Difficulty in breathing

4. Fits of coughing

5. Chemical pneumonia or bronchitis

6. Fluid in the lungs (acute pulmonary edema)

7. Sudden death (AMMONIA only)

B. Signs and symptoms of acute poisoning

1. Irritation of nose and eyes

2. Sneezing

3. Tingling sensation in respiratory tract

4. Coughing

5. Risk of chemical bronchitis

AFTER AN APPARENT ARREST IN THE SYMPTOMS

6. Risk of fluid in the lungs (acute pulmonary edema)

C. *First aid:* see *yellow* pages, section *1*

II. INGESTION

A. Signs and symptoms of acute poisoning

1. Immediate burning sensation in mouth, throat, and stomach. The mucous membranes of the mouth are attacked

2. Pain on swallowing

3. Swelling of the throat

4. Profuse salivation

5. Nausea and vomiting of coffee-grounds-like material (bleeding in stomach)

6. Stomach cramps

7. Rapid breathing

8. State of shock
   - weak and rapid pulse
   - cold sweat—pale complexion
   - lightheadedness
   - cold hands and feet

9. Diarrhea, occasionally blood-stained

10. Risk of stomach perforation with hard, rigid abdomen

B. *First aid:* see *green* pages, section *3*

III. SKIN CONTACT

A. Immediate signs and symptoms

1. Itching and tingling sensation

2. Painful burning

3. Painful ulcerations

4. Skin feels slippery or soapy

5. State of shock
   - weak and rapid pulse

- cold sweat—pale complexion
- lightheadedness
- cold hands and feet

B. *First aid:* see *pink* pages, section *4*

IV. SPLASHING IN EYES

A. Immediate signs and symptoms

1. Immediate, highly painful irritation of eyes and eyelids

2. Intense watering of eyes

3. Victim keeps eyelids tightly closed

4. Burns and irreparable damage to mucous membranes of eyes

5. Serious injury in eyes

B. *First aid:* see *blue* pages, section *2*

Ammonium Carbonate	Potassium Carbonate
Calcium Hydroxide	Sodium Carbonate
Lithium Carbonate	Sodium Silicate
Lithium Hydride	Trisodium Phosphate
Milk of Lime	

In contact with water, the above crystalline substances give alkaline solutions that are highly irritating to mucous membranes of the respiratory and digestive systems and eyes.

LITHIUM CARBONATE is toxic to the kidneys.

I. INHALATION

A. Signs and symptoms of very acute poisoning

1. Irritation of nose, eyes, and throat

2. Sneezing

3. Difficulty in breathing

4. Coughing, occasional fits of coughing

5. Chemical bronchitis

B. Signs and symptoms of acute poisoning

1. Irritation of nose and eyes

2. Tingling sensation in respiratory tract

3. Sneezing

4. Coughing

C. *First aid:* see *yellow* pages, section *2*

II. INGESTION

A. Signs and symptoms of acute poisoning

1. Burning sensation in mouth, chest, and stomach

2. Irritation of mouth

3. Pain in swallowing

4. Stomach cramps

5. Nausea and vomiting (LITHIUM CARBONATE only)

6. Occasional trembling of arms and legs (LITHIUM CARBONATE only)

7. Occasional mental confusion (LITHIUM CARBONATE only)

B. *First aid:* see *green* pages, section *3*

II. CONTACT WITH MOIST SKIN

A. **Immediate signs and symptoms**

1. Itching and tingling sensation

2. Burning sensation

3. Redness and swelling

B. *First aid:* see *pink* pages, section *9*

IV. SPLASHING IN EYES

A. **Immediate signs and symptoms**

1. Mechanical irritation, followed by chemical irritation

2. Pain

3. Watering of eyes

4. Risk of serious injury to eyes if victim keeps eyelids tightly closed

B. *First aid:* see *blue* pages, section *3*, or *blue* pages, section *2*, per **Chemical Index**

Antimony and Compounds
Arsenic
Arsenic Trichloride
Arsenicals
Arsine
Phosphine

Phosphorus
Phosphorus Chlorides
Phosphorus Pentachloride
Phosphorus Trichloride
Stibine

ARSINE, STIBINE, and PHOSPHINE are highly toxic gases.
The liquid chlorides give off highly irritating and toxic fumes.
All the above are toxic to the liver, and kidneys and may cause damage to these organs. They are also toxic to the red blood cells.

I. INHALATION

A. Signs and symptoms of very acute poisoning

1. Passing attack of headache

2. Sudden loss of consciousness

3. Almost instantaneous death

B. Signs and symptoms of acute poisoning

1. Headache

2. Dizziness

3. Nausea, occasionally vomiting

4. Fatigue

5. Difficulty in breathing

6. Pale or BLUISH face

7. Dry cough

8. Cold sweat

9. Abdominal pain with diarrhea

10. Trembling of arms and legs

11. Convulsions

12. Loss of consciousness

13. Risk of fluid in the lungs (pulmonary edema)

14. Death

C. *First aid:* see *yellow* **pages, section** *1*

II. INGESTION

A. Signs and symptoms of acute poisoning

In addition to the severe symptoms given under I. INHALATION, B

1. Irritation of mouth and throat

2. Burning pains in stomach

3. Nausea and vomiting

4. Foul-smelling diarrhea which causes burning pain when passed

B. *First aid:* see *green* **pages, section** *1,* **or** *green* **pages, section** *2,* **per Chemical Index**

III. SKIN CONTACT

These compounds all penetrate or react with healthy skin and cause general poisoning even when there is limited contact with the skin. The chlorides are highly irritating to tissues.

In addition to the general symptoms described under I. INHALATION

A. Immediate signs and symptoms

1. Painful irritation

2. Inflammation

3. Blisters

4. Rapid ulceration of the skin

B. *First aid:* see *pink* pages, section *1*, or see *pink* pages, section *2* (ARSENIC soluble salts only)

IV. SPLASHING IN EYES

A. **Immediate signs and symptoms**

1. Immediate irritation of eyes

2. Watering of eyes

3. Inflammation and burning sensation

4. Risk of serious injury to eyes

B. *First aid:* see *blue* pages, section *1*, or see *blue* pages, section *6* (ARSENIC soluble salts only)

Dibutyltin
Diethyltin
Dihexyltin
Diiododiethyltin
Dimethyltin
Dioctyltin
Tetrabutyltin
Tetraethyltin

Tetraisoalkyltin
Tetrapentyltin
Tetrapropyltin
Tributyltin
Trimethyltin
Triphenyltin
Tripropyltin

The above substances are all toxic to the nervous system and cause long term permanent damage, in addition to damage to the liver and kidneys. Their toxicity increases with decreasing molecular weight.

I. INHALATION

A. Signs and symptoms of very acute poisoning

1. Irritation of nose and eyes

2. Sneezing

3. Headache

4. Nausea

5. General feeling of discomfort

6. Coughing

B. *First aid:* see *yellow* pages, section 5

II. INGESTION

A. Signs and symptoms of acute poisoning

1. Irritation of mouth, throat and stomach

2. Headache

3. Nausea and vomiting

4. Stomach pains

5. Dizziness

6. Convulsions

7. Loss of consciousness

8. Coma

9. Risk of permanent paralysis of the limbs

10. Death

B. *First aid:* **see** *green* **pages, section** *9*

III. SKIN CONTACT

These substances are irritating to the skin in varying degrees.

A. Immediate symptoms

1. Itching

2. Irritation

3. Redness and swelling

4. Blisters may occur, followed by ulceration of the skin

B. *First aid:* **see** *pink* **pages, section** *2*

IV. SPLASHING IN EYES

Vapors of these substances are irritating to the eyes.

A. Immediate symptoms

1. Irritation and itching

2. Watering of eyes

3. Redness and swelling of eyelids

B. *First aid:* **see** *blue* **pages, section** *6*

Aldicarb	OMPA
Carbamates	Organophosphate Compounds
Chlorthion	Paraoxon
DDVP	Parathion
Demeton	Phorate
Diazinon	Phosdrin
Dipterex	Phosphoric Ester
EPN	Ronnel
Isopestox	Sulfotepp
Leptophos	TEPP
Malathion	Trithion
Methyl Parathion	

IMPORTANT NOTE: A splash of a single drop of PARATHION in the eye may result in serious effects, and even death.

The ORGANOPHOSPHORUS and CARBAMATE COMPOUNDS are volatile, extremely toxic compounds, usually with very rapid effects. They can be absorbed by skin or eye contact as well as by ingestion or inhalation. In some cases, signs and symptoms may be delayed up to 4 hours after exposure. ALWAYS send the victim to a medical facility for immediate follow-up.

They block the action of cholinesterase in the tissues which results in widespread nervous system disorders leading to serious muscular disturbances and increased secretions of saliva, sweat and phlegm in the lungs.

Atropine given intravenously by medical personnel is life saving.

I. INHALATION

A. Signs and symptoms of acute poisoning

1. Profuse sweating

2. Slow heart beat

3. Nausea and vomiting

4. Salivation and watering eyes

5. Stomach cramps

6. Muscle twitching

7. Diarrhea

8. Weakness

9. Headache and dizziness

10. Trouble breathing with wheezing

11. Very small pupils

12. Blurred vision

13. Anxiety and agitation

14. Bizarre behavior

15. Cough producing sputum

16. Hypersecretion in the lungs

17. Weakness of respiratory muscles

18. Fluid in the lungs (pulmonary edema)

19. Slow pulse leading eventually to stoppage of heart

20. Convulsions

21. High fever

22. Incontinence of bladder and bowels

23. Loss of consciousness

24. Death

B. *First aid:* see *yellow* **pages, section** *3*

II. INGESTION

A. Signs and symptoms of acute poisoning

See symptoms described under I. INHALATION

B. *First aid:* see *green* **pages, section** *2*

III. SKIN CONTACT

A. Immediate signs and symptoms

See symptoms described under I. INHALATION

B. *First aid:* see *pink* pages, section *11*

IV. SPLASHING IN EYES

The ORGANOPHOSPHATE and CARBAMATE COMPOUNDS are rapidly absorbed by mucous membranes of the eyes and can cause generalized poisoning within a short period of time.

A. Immediate signs and symptoms

1. Irritation

2. Watering of eyes

3. Blurred vision

These are quickly followed by the symptoms described under I. IN-HALATION

B. *First aid:* see *blue* pages, section *9*

Lead (dust and fumes)	Lead Dioxide
Lead Acetate	Lead Nitrate
Lead Antimonate	Lead Oxide (PbO)
Lead Arsenate	Lead Oxide (red)
Lead Carbonate	Lead Oxychloride
Lead Chromate	Lead Subacetate
Lead Chromate (yellow)	Lead Sulfide

These substances are toxic to the kidneys, nerves, muscles, and blood. The water-soluble salts and the oxides react with gastric juices (HCl) and cause gastrointestinal irritation and lead poisoning.

As a result of inhalation, ingestion, and skin contact, these substances may eventually cause chronic lead poisoning.

I. INHALATION

Poisoning may be caused when these compounds are in the form of fine dust as well as by their decomposition products during heating or cleaning.

A. Signs and symptoms of acute poisoning

1. Slight irritation of nose and eyes

2. Headache

3. Stomach cramps

4. Fatigue

5. Seizures

6. Confusion

7. Coma

8. Death

B. *First aid:* see *yellow* pages, Section *6*

II. INGESTION

A. Signs and symptoms of acute poisoning

1. Metallic taste in mouth

2. Constriction of throat

3. Stomach pain

4. Nausea and vomiting, occasionally of blood

5. Cramping abdominal pain

6. Diarrhea

7. Seizures

8. Coma

9. Death

B. *First aid:* see *green* pages, section *2*

III. SKIN CONTACT

A. Immediate signs and symptoms

1. Irritation

2. Inflammation

B. *First aid:* see *pink* pages, section *2*

IV. SPLASHING IN EYES

The insoluble salts of lead are mechanical irritants whereas the soluble salts are chemical irritants to the eyes.

A. Immediate signs and symptoms

1. Irritation and redness

2. Watering of eyes

3. Inflammation of eyelids

B. *First aid:* see *blue* pages, section *6*

Acetone Cyanohydrin	Ferrocyanides
Acetonitrile	Hydrocyanic Acid
Acrylonitrile	Isobutyronitrile
Adiponitrile	Malononitrile
Bitter Almond Oil (Amygdalin)	Methacrylonitrile
Cherry Laurel Water	Nitroferricyanides (salts)
Cyanogen Bromide	Potassium Cyanide
Cyanogen Chloride	Sodium Cyanide
Cyanogen Iodide	Tetramethyl Succinonitrile
Ferricyanides	

HYDROCYANIC ACID, POTASSIUM CYANIDE, and SODIUM CY-ANIDE are fast-acting, highly toxic substances that interfere with the use of oxygen by living cells. This can be the result of either inhalation or ingestion.

ACRYLONITRILE is toxic when inhaled and highly irritating to skin and eyes.

The FERRICYANIDES and FERROCYANIDES are less toxic.

I. INHALATION

A. Symptoms of very acute poisoning

1. Victim cries out before losing consciousness

2. Victim falls to the ground

3. Wheezing

4. Foaming at mouth

5. Violent convulsions

6. Almost immediate death

B. Signs and symptoms of acute poisoning

1. Excitement phase
   headache
   breath smells of bitter almond
   dizziness
   nausea, occasionally vomiting

rapid breathing
anxiety and excitement

2. Depression phase
difficulty in breathing
chest pain
drowsiness

3. Convulsion phase
convulsions
jaws clenched together
foaming at mouth
loss of consciousness

4. Paralysis phase
deep coma
dilated pupils
weak and irregular pulse
breathing stops
death

If the subject survives, there is risk of permanent nervous system damage.

C. Signs and symptoms of slight poisoning

1. Headache

2. Dizziness

3. Anxiety

4. Difficulty in breathing

D. *First aid:* see *yellow* pages, section *8*

II. INGESTION

A. Signs and symptoms

See symptoms described under I. INHALATION, B and C

1. Burning tongue and mouth

2. Salivation

3. Nausea

B. *First aid:* **see** *green* **pages, section** *11*

III. SKIN CONTACT

A. Signs and symptoms

The gaseous and liquid compounds are quickly absorbed by the skin and cause symptoms described under I. INHALATION, B and C.
Depending on their nature, they can be very or only slightly irritating.

1. Large blisters (ACRYLONITRILE)

B. *First aid:* **see** *pink* **pages, section** *8*

IV. SPLASHING IN EYES

When absorbed by mucous membranes of the eyes, these compounds can cause the same symptoms described in I. INHALATION, C.

A. Immediate signs and symptoms

1. Irritation

2. Watering of eyes

B. *First aid:* **see** *blue* **pages, section** *5*

Chlorophenoxy Compounds	Nitroglycerin
DNBP	Nitrocresolic Herbicides
DNOC	Nitrophenolic Herbicides
Dinitrocresols	Nitrophenols
Dinitrophenols	Pentachlorophenate
Ethyl Nitrate	Pentachlorophenol
Ethylene Glycol Dinitrate	Propyl Nitrate
Methyl Nitrate	

These nitro-organic or chlorophenolic compounds may inhibit phosphate bonds and consequently interfere with the way that cells use oxygen, causing high fever.

They are irritating to lungs, skin, and eyes and may cause damage in liver and kidneys.

I. INHALATION

A. Signs and symptoms of acute poisoning

1. Irritation of nose

2. Headache

3. Coughing

4. Profuse sweating

5. Thirst

6. Intense fatigue

7. High fever

8. Rapid pulse

9. Warm flushed skin

10. Rapid, difficult breathing

11. Anxiety and confusion

12. BLUISH face and lips; YELLOW in the case of DINITROCRESOLS only

13. Convulsions

14. Loss of consciousness

15. Risk of fluid in lungs (severe pulmonary edema) if victim survives

16. Death due to heart failure

B. *First aid:* **see** *yellow* **pages, section** *4*

II. INGESTION

A. Signs and symptoms of acute poisoning

1. Burning sensation in mouth and throat

2. Salivation

3. Dizziness

4. Headache

5. Nausea and vomiting

6. High fever with warm, flushed skin

7. Profuse sweating

8. Thirst

9. Rapid and difficult breathing

10. BLUISH face and lips; YELLOW in the case of DNOC only

11. Rapid pulse

12. Intense fatigue

13. Anxiety and confusion

14. Convulsions

15. Loss of consciousness

16. Risk of fluid in lungs (pulmonary edema) if victim survives

17. Death

B. *First aid:* **see** *green* **pages, section** *7*

III. SKIN CONTACT

These substances can all be absorbed through healthy skin, especially when they are dissolved in organic solvents. They may cause delayed symptoms identical to those described under I. INHALATION.

A. Immediate signs and symptoms

1. Irritation

2. Redness and swelling

3. Blisters

B. Delayed signs and symptoms

See I. INHALATION

C. *First aid:* see *pink* pages, section *6*

IV. SPLASHING IN EYES

A. Immediate signs and symptoms

1. Burning pain in eyes

2. Irritation

3. Watering of eyes

4. Inflammation of eyelids

B. *First aid:* see *blue* pages, section *3*

27

Disodium Phosphate
Magnesium Chloride
Magnesium Sulfate

Sodium Sulfate
Sodium Thiocyanate
Sodium Thiosulfate

These compounds are crystalline, water-soluble, and considered nontoxic and nonirritating to skin and mucous membranes.

Except for SODIUM THIOCYANATE, they are cathartics and consequently used as first aid to accelerate the elimination of toxic products from the intestines. MAGNESIUM SULFATE and SODIUM SULFATE are most frequently used.

I. INHALATION

A. Immediate signs and symptoms

1. Mild nose irritation with sneezing

B. *First aid:* see *yellow* **pages, section** *6*

II. INGESTION

A. Signs and symptoms

1. Diarrhea

B. First aid: none needed

III. SKIN CONTACT

Although these substances are not irritating to skin, it is advisable to avoid prolonged contact with them in powder form or concentrated solutions.

A. *First aid:* see *pink* **pages, section** *7,* **or see** *pink* **pages, section** *8* **(THIOCYANATES only)**

IV. SPLASHING IN EYES

A. Immediate signs and symptoms

1. Mechanical irritation

2. Watering of eyes

B. *First aid:* **see** *blue* **pages, section** *6,* **or see** *blue* **pages, section** *5* **(SODIUM THIOCYANATE only)**

Aluminum Alkyls	Triethylaluminum
Diethylaluminum Chloride	Triisobutylaluminum
Diethylaluminum Hydride	Trimethylaluminum

These substances occur as liquids at room temperatures. They are generally diluted with a hydrocarbon-type solvent and kept in a nitrogen atmosphere according to very strict safety regulations.

These compounds burst spontaneously and almost instantaneously into flames as soon as they come into contact with atmospheric oxygen or water and give off nontoxic fumes of ALUMINUM OXIDE (see **white** pages, section **14**).

I. INHALATION OF FUMES

A. Signs and symptoms of acute inhalation

1. Sneezing

2. Slight irritation of nose

B. *First aid:* see *yellow* pages, section *6*

II. INGESTION

Unlikely because of strict handling requirements.

III. SKIN CONTACT

Unlikely because of the protective equipment required during handling of the product. However, if ignition of the product occurs:

A. Immediate signs and symptoms

1. Third-degree burns (irreparable damage to tissues)

2. State of shock
   - weak and rapid pulse
   - cold sweat—pale complexion

- lightheadedness
- cold hands and feet

3. Loss of consciousness

4. Death

B. *First aid:* **see** *pink* **pages, section** *10*

IV. SPLASHING IN EYES

Unlikely because of the protective equipment required during handling of the product.

A. Immediate signs and symptoms

1. Burning, by flames, of face and eyelids

2. Excruciating pain

3. Swelling of eyelids

4. Possible injury of eyes and impairment of vision

B. *First aid:* **see** *blue* **pages, section** *8*

29

Alkanes (liquids/solids)
Asphalt Fumes
Chloronaphthalenes
Diphenyl
Diphenylamine

Hexachlorobenzene
Paraffins
Polybrominated Biphenyls (PBBs)
Polychlorinated Biphenyls (PCBs)

These substances are oily liquids and waxy solids. They are soluble in organic solvents. They are considered to present little risk on acute exposure

I. INHALATION

A. Signs and symptoms of severe poisoning

1. Irritation of nose

2. Headache

3. Nausea may occur

4. Coughing

B. *First aid:* see *yellow* **pages, section** *2*

II. INGESTION

Only the amines in the list above are slightly irritating to mucous membranes of the digestive system.

A. Signs and symptoms of severe poisoning

1. Slight irritation of stomach

2. Nausea, occasionally vomiting

3. General fatigue

B. *First aid:* see *green* **pages, section** *10*

III. SKIN CONTACT

In solution in organic solvents, these substances will slowly penetrate the skin. The immediate symptoms that can be observed are the result of poisoning by the solvent used.

A. *First aid:* see *pink* pages, section *6,* or *pink* pages, section *9,* per Chemical Index

IV. SPLASHING IN EYES

A. Immediate signs and symptoms

1. Mechanical irritation
2. Watering of eyes

B. *First aid:* see *blue* pages, section *6*

30

Dipyridyl Chloride   Gramoxone
Dipyridyl Dimethyl Sulfate   Paraquat
Diquat   Quaternary Ammonium Compounds

The pesticides belonging to the dipyridyl group are not volatile. Their solutions are highly toxic when inhaled, and particularly when ingested, while resorption through healthy skin is relatively slow.

They are irritating to eyes, and toxic to liver, kidneys, and lungs.

I. INHALATION

A. Signs and symptoms of acute poisoning

1. Irritation of nose and eyes

2. Sneezing

3. Nose bleed

4. Painful obstruction of throat

B. *First aid:* **see** *yellow* **pages, section** *4*

II. INGESTION

A. Signs and symptoms of acute poisoning

1. Burning sensation in mouth and throat

2. Irritation of mucous membranes

3. Difficulty in swallowing

4. Burning sensation in stomach

5. Nausea and vomiting, possibly of blood

6. Painful abdominal cramps

7. Diarrhea, may be blood-stained

8. There is risk of serious lesions in

- the liver
- the kidneys
- the lungs (fatal chemical pneumonia)

B. *First aid:* see *green* pages, section *2,* or *green* pages, section *5,* per **Chemical Index**

III. SKIN CONTACT

A. Immediate symptoms

1. Irritation

2. Inflammation

3. Painful blisters

B. *First aid:* see *pink* pages, section *6*

IV. SPLASHING IN EYES

A. Immediate symptoms

1. Painful burning sensation

2. Watering of eyes

3. Inflammation

4. Sensitivity to light

B. *First aid:* see *blue* pages, section *5*

Aldrin	Dimethylhydrazine (1,1-)
Aminopyridine	Hydrazine
Camphor	Lindane
Chlordane	Organochlorines
Decaborane	Pentaborane

These pesticides and chemicals can be absorbed through intact skin as well as by inhalation and ingestion. Their major target is the nervous system and convulsions may be the first symptom.

I. INHALATION

A. Signs and symptoms of acute poisoning

1. Apprehension and excitability

2. Dizziness

3. Headache

4. Numbness and weakness in limbs

5. Muscle twitching

6. Convulsions

7. Shallow, slow respirations

8. Pallor

9. BLUISH face

10. Respirations cease

11. Death

B. *First aid:* see *yellow* pages, section *5*, or *yellow* pages, section *4* per **Chemical Index.**

II. INGESTION

A. Signs and symptoms of acute poisoning

1. Nausea and vomiting

2. See signs and symptoms in I. INHALATION

B. *First aid:* see *green* pages, section 5

III. SKIN CONTACT

A. Immediate signs and symptoms

1. Apprehension and anxiety
2. Tremors
3. Confusion
4. Seizures
5. Slow or shallow breathing
6. BLUISH face

B. *First aid:* see *pink* pages, section 2, or *pink* pages, section 6, per **Chemical Index**

IV. EYE CONTACT

A. Immediate signs and symptoms

1. Irritation of eyes and eyelids
2. See signs and symptoms in section III. SKIN CONTACT

B. *First aid:* see *blue* pages, section 5

First Aid in case of Inhalation of:

Acetaldehyde	Hydrogen Chloride
Acetic Acid	Hydrogen Peroxide
Acetic Anhydride	Hydrogen Selenide
Acrolein	Hydrogen Sulfide
Aliphatic Amines	Iodine
Ammonia	Isobutyraldehyde
Ammonium Hydroxide	Isopropylamine
Ammonium Sulfide	Maleic Anhydride
Antimony and Compounds	Methyl Chloroformate
Arsenic	Methyl Mercaptan
Arsenic Trichloride	Methylamine
Arsenicals	Nitric Acid
Arsine	Nitric Oxide
Benzyl Chloride	Nitrogen Dioxide
Boron Trifluoride	Nitrogen Trifluoride
Bromine	Osmic Acid
Butylamine	Oxygen Difluoride
Butyraldehyde	Ozone
Carbon Disulfide	Peracetic Acid
Chlorine	Perchloric Acid
Chlorine Dioxide	Perchloryl Fluoride
Chlorine Trifluoride	Phenol
Chloroacetaldehyde	Phenylenediamine (p-)
Chloroacetic Acid	Phosgene
Creosote	Phosphine
Cresols	Phosphoric Acid
Crotonaldehyde	Phosphorus
Dibutylamine	Phosphorus Chlorides
Dichloro-5,5-Dimethylhydantoin	Phosphorus Pentachloride
Diethylamine	Phosphorus Pentasulfide
Dimethyl Sulfate	Phosphorus Trichloride
Dimethylamine	Potassium Chlorite
Dipropylamine	Potassium Fluoride
Ethanolamine	Propionaldehyde
Ethyl Chloroformate	Propylamine
Ethylamine	Quinone
Fluorine	Resorcinol
Fluosilicic Acid	Selenium Hexafluoride
Formaldehyde	Silane
Formic Acid	Sodium Chlorite
Hydriodic Acid	Sodium Fluoride
Hydrochloric Acid	Stibine
Hydrofluoric Acid	Sulfur Dioxide
Hydrogen Bromide	Sulfur Trioxide

Sulfuric Acid	Trichloroacetic Acid
Sulfurous Acid	Triethylamine
Tellurium Hexafluoride	Trimethylamine
Tributyl Phosphate	

---

1. Remove the victim from the contaminated area while protecting yourself from exposure by wearing an appropriate respirator. Put a similar respirator on the victim.

2. Have someone call the Emergency Medical Service and arrange for transport to a medical facility. Tell them the nature of the exposure.

A. If the victim has stopped breathing:

1. Open his airway, loosen his collar and belt, and administer artificial respiration using a bag-valve mask or the chest pressure-arm lift technique.

2. Administer oxygen through the bag-valve mask or by an oxygen mask if using the chest pressure-arm lift technique.

3. Check the pulse. If the heart stops, administer CPR.

4. Continue your efforts until help arrives or the victim starts to breathe on his own. Do not leave him alone.

5. Keep the victim warm and quiet.

B. If the victim is unconscious but breathing:

1. Lay him on his back. If he is vomiting, turn his head to the side.

2. Clear his airway and loosen tight clothing.

3. If available, give him oxygen to breathe.

4. Keep him warm and quiet.

5. Do not leave him unattended.

6. Do not give an unconscious person anything to drink.

C. If the victim is conscious but coughing or short of breath:

1. Lay him down, cover him with a blanket and keep him quiet.

2. Loosen tight clothing.

3. Give him oxygen to breathe until help arrives.

4. He may have water to sip.

D. If the victim is conscious but not coughing or short of breath:

1. Lay him down, cover him with a blanket and keep him quiet.

2. Loosen tight clothing.

Alkanes (liquids/solids)
Ammonium Carbonate
Ammonium Chlorate
Ammonium Perchlorate
Asphalt Fumes
Barium (soluble salts)
Barium Acetate
Barium Carbonate
Barium Chloride
Barium Fluoride
Barium Hydroxide
Barium Nitrate
Barium Oxide
Barium Sulfide
Calcium Hydroxide
Chloronaphthalenes
Diphenyl

Diphenylamine
Hexachlorobenzene
Lithium Carbonate
Lithium Hydride
Oxalic Acid
Paraffins
Polybrominated Biphenyls (PBBs)
Polychlorinated Biphenyls (PCBs)
Potassium Carbonate
Potassium Chlorate
Potassium Perchlorate
Sodium Carbonate
Sodium Chlorate
Sodium Perchlorate
Sodium Silicate
Trisodium Phosphate

1. Remove the victim from the contaminated area while protecting yourself from exposure by wearing an appropriate respirator. Put a similar respirator on the victim.

A. If the victim is coughing or short of breath:

1. Loosen tight clothing.

2. Encourage him to spit out what he coughs up.

3. Give him oxygen to breathe until help arrives.

4. Arrange to have him taken to a medical facility for examination and treatment.

B. If the victim is not coughing but feels irritation in his nose:

1. Encourage him to blow his nose to remove the substance.

2. Arrange to have him taken to a medical facility for examination and treatment.

Acetone
Acetylene
Aldicarb
Aliphatic Alcohols—Amyl
Aliphatic Alcohols—Butyl
Alkanes (gasses, C_1 to C_4)
Amyl Acetate
Benzene
Bromoform
Butadiene
Butane
Butanol
Butyl Acetate
Butyltoluene
Carbamates
Carbon Dioxide
Carbon Dioxide Snow
Carbon Monoxide
Carbon Tetrachloride
Chlorobenzene
Chlorobromomethane
Chlorodifluoroethane
Chlorodifluoromethane
Chloroethane
Chlorofluoroethane
Chlorofluoromethane
Chloroform
Chloromethane
Chloropentafluoroethane
Chloropropane
Chloropropene
Chlorotrifluoroethylene
Chlorotrifluoromethane
Chlorthion
Cumene
Cyclohexane
DDVP
Decane
Decanol
Demeton
Diacetone Alcohol
Diazinon
Dichlorobenzene
Dichlorodifluoromethane

Dichloroethane
Dichloroethylene
Dichlorofluoromethane
Dichloropropane
Dichlorotetrafluoroethane
Diethylene Glycol
Difluoroethanes
Difluoroethylene
Diisobutylcarbinol
Dioxane
Dipterex
EPN
Ethane
Ethyl Acetate
Ethyl Alcohol
Ethyl Ether
Ethyl Fluoride
Ethylbenzene
Ethylene
Ethylene Chlorohydrin
Ethylene Dichloride
Ethylene Glycol
Ethylene Glycol Monomethyl Ether
Ethylhexyl Acetate
Fluoromethane
Freon 11, 12, 13, 14, 21, 22, 112, 113, 114, 115, 116, 142b, 143, 151a, 152a
Furfuryl Alcohol
Gasoline
Glycerin
Halothane
Heptane
Heptanol
Hexachloroethane
Hexafluoroethane
Hexane
Hexanol
Isobutyl Acetate
Isopestox
Isopropyl Acetate
Isopropyl Alcohol
Leptophos
Liquefied Petroleum Gas

Malathion
Methane
Methyl Acetate
Methyl Alcohol
Methyl n-Butyl Ketone
Methyl Chloride
Methyl Ethyl Ketone
Methyl Isoamyl Ketone
Methyl Isobutyl Ketone
Methyl Isopropyl Ketone
Methyl Parathion
Methylchloroform
Methylene Chloride
Methylene Fluoride
Naptha
Nitrogen
Nitromethane
Nonane
OMPA
Octane
Organophosphate Compounds
Paraoxon
Parathion
Pentachloroethane
Pentane
Pentanol
Petroleum Ethers
Phorate
Phosdrin
Phosphoric Ester

Propane
Propyl Acetate
Propyl Alcohol
Propylene
Propylene Glycol
Ronnel
Stoddard Solvent
Sulfotepp
TEPP
Tetrachlorodifluoroethane
Tetrachloroethane
Tetrachloroethylene
Tetrafluoroethylene
Tetrafluoromethane
Toluene
Trichloroethane
Trichloroethylene
Trichlorofluoromethane
Trichlorotrifluoroethane
Triethylene Glycol
Trifluoroethane
Trifluoromethane
Trithion
Turpentine
Vinyl Acetate
Vinyl Chloride
Vinyl Fluoride
Vinylidene Chloride
Vinylidene Fluoride
Xylene

---

1. Remove the victim from the contaminated area while protecting yourself from exposure by wearing an appropriate respirator. Put a similar respirator on the victim.

2. Have someone call the Emergency Medical Service and arrange for transport to a medical facility.

3. Remove contaminated clothing and equipment, while wearing gloves, being careful not to contaminate yourself.

A. If the victim has stopped breathing:

1. Open his airway, loosen his collar and belt, and administer artificial respiration using a bag-valve mask or mouth to mouth resuscitation.

2. Administer oxygen through the bag-valve mask.

3. Check the pulse. If the heart stops, administer CPR. If the heart beat is very slow, irregular or weak, be prepared to administer CPR.

4. Continue your efforts until help arrives or the victim starts to breathe on his own. Do not leave him alone.

5. Keep the victim warm and quiet.

B. If the victim is unconscious but breathing:

1. Lay him on his back. If he is vomiting, turn his head to the side.

2. Clear and open his airway and loosen tight clothing.

3. If available, give him oxygen to breathe.

4. Keep him warm and quiet. Check his pulse periodically and be ready to administer CPR.

5. Do not leave him unattended.

6. Do not give an unconscious person anything to drink.

C. If the victim is conscious but coughing or short of breath:

1. Lay him down, cover him with a blanket and keep him quiet.

2. Loosen tight clothing.

3. Give him oxygen to breathe until help arrives.

4. Check his pulse periodically. Do not leave him unattended.

D. If the victim is conscious and not coughing or short of breath:

1. Lay him down, cover him with a blanket and keep him quiet.

2. Loosen tight clothing.

3. Check his pulse periodically. Do not leave him unattended.

Acrylamide
Allyl Alcohol
Allyl Chloride
Allyl Glycidyl Ether
Allyl Propyl Disulfide
Aniline
Anisidines (ortho)
Anisidines (para)
Benzidine
Bis(Chloromethyl) Ether
Butyl Glycidyl Ether (n-)
Chloro-1-Nitropropane (1-)
Chloroacetophenone (2-)
Chlorobenzylidene Malonitrile
Chlorophenoxy Compounds
Chloropicrin
Cyclohexanol
Cyclohexanone
DNBP
DNOC
Diazomethane
Diborane
Dibutyl Phthalate
Diepoxybutane
Diethylaminoethanol
Diglycidyl Ether
Dimethylaniline
Dinitrobenzene
Dinitrocresols
Dinitrophenols
Dinitrotoluene
Dipyridyl Chloride
Dipyridyl Dimethyl Sulfate
Diquat
Epichlorohydrin
Ethyl Acrylate
Ethyl Nitrate
Ethylene Glycol Dinitrate
Ethylene Oxide
Ethyleneimine
Furfural
Glutaraldehyde
Glycidol

Glycidyl Acrylate
Gramoxone
Hydrazine
Hydroquinone
Ketene
Methyl Acrylate
Methyl Bromide
Methyl Isocyanate
Methyl Methacrylate Monomer
Methyl Nitrate
Methylenebis (Phenyl Isocyanate)
Monomethylhydrazine
Naphthalene
Naphthylamines
Nickel (fumes and dust)
Nickel Carbonyl
Nitroanilines
Nitrobenzene
Nitrochlorobenzene (p-)
Nitroglycerin
Nitrocresolic Herbicides
Nitrophenolic Herbicides
Nitrophenols
Nitrotoluene
Paraquat
Pentachlorophenate
Pentachlorophenol
Perchloromethyl Mercaptan
Phenylhydrazine
Phenylhydroxylamine
Phenylnaphthylamine
Picric Acid
Propyl Nitrate
Propylene Glycol Monomethyl Ether
Propylene Oxide
Pyrethrins
Quaternary Ammonium Compounds
Styrene
Tetranitromethane
Tetryl
Tolidine (o-)
Toluene, 2,4-di-Isocyanate
Toluene, 2,6-di-Isocyanate

Toluidine   Trinitrotoluene
Trinitrobenzene   Xylidine

---

1. Remove the victim from the contaminated area while protecting yourself from exposure by wearing an appropriate respirator. Put a similar respirator on the victim.

2. Have someone call the Emergency Medical Service and arrange for transport to a medical facility.

3. Remove contaminated clothing and equipment using gloves and being careful not to contaminate yourself.

A. If the victim has stopped breathing:

1. Open his airway, loosen his collar and belt, and administer artificial respiration using a bag-valve mask or mouth to mouth resuscitation.

2. Administer oxygen through the bag-valve mask.

3. Check the pulse. If the heart stops, administer CPR.

4. Continue your efforts until help arrives or the victim starts to breathe on his own. Do not leave him alone.

B. If the victim is unconscious but breathing:

1. Lay him on his back and turn his head to the side.

2. Clear his airway and loosen tight clothing.

3. If available, give him oxygen to breathe.

4. Keep him warm and quiet.

5. Do not leave him unattended.

6. Do not give an unconscious person anything to drink.

C. If the victim is conscious but appears BLUE or BROWNISH or is short of breath:

1. Lay him down, cover him with a blanket and keep him quiet.

2. Loosen tight clothing.

3. Give him oxygen to breathe until help arrives.

5

First Aid in Case of Inhalation of:

Aldrin	Mercury (soluble salts)
Aminopyridine	Mercury Acetate
Camphor	Mercury Fulminate
Chlordane	Mercury Nitrate
Decaborane	Mercury Oxycyanide
Dibutyllead	Methylmercury
Dibutyltin	Methylmercury Borate
Diethyllead	Methylmercury Hydroxide
Diethylmercury	Methylmercury Iodide
Diethyltin	Methylmercury Nitrate
Dihexyltin	Methylmercury Phosphate
Diiododiethyltin	Organochlorines
Dimethylhydrazine (1,1-)	Pentaborane
Dimethylmercury	Phenylmercuric Acetate
Dimethyltin	Phenylmercury
Dioctyltin	Phenylmercury Oleate
Ethylmercuric Chloride	Tetrabutyltin
Ethylmercuric Hydroxide	Tetraethyllead
Ethylmercury	Tetraethyltin
Lead Oleate	Tetraisoalkyltin
Lead Phenate	Tetramethyllead
Lead Phthalate	Tetrapentyltin
Lead Stearate	Tetrapropyltin
Lindane	Tributyllead
Mercuric Chloride	Tributyltin
Mercuric Iodide (red)	Triethyllead
Mercurous Chloride	Trimethyllead
Mercurous Iodide	Trimethyltin
Mercury (metal)	Triphenyltin
Mercury (organic compounds)	Tripropyltin

1. Remove the victim from the contaminated area while protecting yourself from exposure by wearing an appropriate respirator. Put a similar respirator on the victim.

2. Have someone call the Emergency Medical Service and arrange for transport to a medical facility, even if there are no symptoms.

3. Remove contaminated clothing and equipment using gloves and being careful not to contaminate yourself.

4. Lay him down, cover him with a blanket and keep him quiet.

5. Loosen tight clothing.

6. If the victim is coughing or is short of breath, give him oxygen to breathe until help arrives.

7. Observe for seizures. Do not leave the victim alone.

First Aid in Case of Inhalation of:

Aluminum (dust)
Aluminum Alkyls
Aluminum Hydrate
Aluminum Hydroxide
Aluminum Oxide
Asbestos
Calcium Carbonate
Calcium Chloride
Carbon
Carbon Black
Diethylaluminum Chloride
Diethylaluminum Hydride
Disodium Phosphate
Dithiocarbamates
Fibrous Glass
Kaolin
Lead (dust and fumes)
Lead Acetate
Lead Antimonate
Lead Arsenate
Lead Carbonate
Lead Chromate
Lead Chromate (yellow)
Lead Dioxide
Lead Nitrate

Lead Oxide (PbO)
Lead Oxide (red)
Lead Oxychloride
Lead Subacetate
Lead Sulfide
Magnesium Chloride
Magnesium Sulfate
Polyvinyl Chloride
Potassium Chloride
Silica
Sodium Bicarbonate
Sodium Chloride
Sodium Sulfate
Sodium Thiocyanate
Sodium Thiosulfate
Talc
Thiocarbamates
Titanium (dust and fumes)
Titanium Dioxide
Triethylaluminum
Triisobutylaluminum
Trimethylaluminum
Tungsten Carbide
Yttrium and Compounds

1. Remove the victim from the contaminated area while protecting yourself, if necessary, from exposure by wearing an appropriate respirator. Put a similar respirator on the victim. If no respirators are available, remove the victim as quickly as possible.

2. Remove contaminated clothing and equipment preferably while wearing gloves and respirator. Remove after wetting down if possible.

3. Encourage the victim to cough, spit out and blow his nose to remove dust.

4. Consult a physician for follow up.

7

First Aid in Case of Inhalation of:

Alkali Dichromates
Alkali Meta-Borates
Aluminum Chloride
Aluminum Trichloride
Boric Acid
Cadmium (dust and fumes) (metal)
Calcium Carbide
Calcium Dichromate
Calcium Hypochlorite
Calcium Oxide
Caprolactam
Cement
Chlorinated Lime
Chromic Acid
Chromium Chloride
Copper Chloride
Copper Sulfate
Iron Chloride
Lime
Perborates
Phthalic Anhydride

Platinum and Compounds
Potassium
Potassium Chromate
Potassium Dichromate
Potassium Fluosilicate
Potassium Hydroxide
Potassium Oxide
Sodium
Sodium Borate
Sodium Chromate
Sodium Dichromate
Sodium Fluosilicate
Sodium Hydroxide
Sodium Hypochlorite
Sodium Oxide
Sodium Peroxide
Titanium Chlorides
Trimellitic Anhydride
Uranium and Compounds
Vanadium and Compounds
Zinc Chloride

1. Remove the victim from the contaminated area while protecting yourself, if necessary, from exposure by wearing an appropriate respirator. Put a similar respirator on the victim, if necessary.

2. Have someone call the Emergency Medical Service and arrange for transport to a medical facility.

A. If the victim is unconscious but breathing:

1. Lay him on his back and turn his head to the side.

2. Clear his airway and loosen tight clothing.

3. If available, give him oxygen to breathe until help arrives.

4. Keep him warm and quiet.

5. Do not leave him unattended.

6. Never give an unconscious person anything to drink.

B. If the victim is conscious but his face is BLUISH, or he is coughing or short of breath:

1. Lay him down with his head and chest propped up.

2. Cover him with a blanket and keep him quiet.

3. Loosen tight clothing.

4. Encourage him to blow his nose and to cough up and spit out.

5. Give him oxygen to breathe until help arrives.

6. He may have water to sip.

7. Do not leave him unattended.

C. If the victim is conscious, and breathing easily:

1. Lay him down, cover him with a blanket and keep him quiet.

2. Loosen tight clothing.

3. Make sure that he sees a physician, even if he has no immediate symptoms.

8

First Aid in Case of Inhalation of:

Acetone Cyanohydrin	Ferrocyanides
Acetonitrile	Hydrocyanic Acid
Acrylonitrile	Isobutyronitrile
Adiponitrile	Malononitrile
Bitter Almond Oil (Amygdalin)	Methacrylonitrile
Cyanogen Bromide	Nitroferricyanides (salts)
Cyanogen Chloride	Potassium Cyanide
Cyanogen Iodide	Sodium Cyanide
Ferricyanides	Tetramethyl Succinonitrile

1. Remove the victim from the contaminated area only after protecting your-self from exposure by wearing an appropriate respirator and occlusive clothing. Put a similar respirator on the victim.

2. Have someone call the Emergency Medical Service and arrange for trans-port to a medical facility. Inform them of the nature of the exposure.

3. Remove contaminated clothing and equipment while wearing gloves and a respirator.

A. If the victim has stopped breathing:

1. Open his airway, loosen his collar and belt, and administer artificial res-piration using a bag-valve mask or the chest pressure-arm lift technique.

2. If available, break a perle of amyl nitrite in a handkerchief and hold it over the victim's nose or place it over his nose under the mask while continuing to ventilate. Remove the perle for 30 seconds and then replace it for 30 seconds. Use a fresh perle every 5 minutes until 3 or 4 perles have been used.

3. Check the pulse. If the heart stops, administer CPR.

4. Continue your efforts until help arrives or the victim starts to breathe on his own.

5. Administer oxygen by mask.

6. Keep the victim warm and quiet.

B. If the victim is unconscious but breathing:

1. Lay him on his back. If he is vomiting, turn his head to the side.

2. Clear his airway and loosen tight clothing.

3. If available, break a perle of amyl nitrite in a handkerchief and hold under the victim's nose for 30 seconds, then remove for 30 seconds. Break a new perle every 5 minutes until 3 or 4 are used up.

4. If available, give him oxygen to breathe after amyl nitrite is administered.

5. Keep him warm and quiet.

6. Do not leave him unattended.

7. Never give an unconscious person anything to drink.

C. If the victim is conscious:

1. If available, break a perle of amyl nitrite in a handkerchief and hold under the victim's nose for 30 seconds, then remove for 30 seconds. Break a new perle every 5 minutes until 3 or 4 are used up.

2. Lay him down, cover him with a blanket and keep him quiet.

3. Loosen tight clothing.

4. Give him oxygen to breathe after amyl nitrite has been administered.

First Aid in Case of Ingestion of:

Acetic Acid	Nitric Acid
Acetic Anhydride	Nitrogen Dioxide
Arsenic	Osmic Acid
Arsenic Trichloride	Oxalic Acid
Arsenicals	Peracetic Acid
Barium Fluoride	Perchloric Acid
Bromine	Phenol
Butyraldehyde	Phosphoric Acid
Calcium Hypochlorite	Phosphorus
Chlorine	Phosphorus Chlorides
Chlorine Dioxide	Phosphorus Pentachloride
Chlorine Trifluoride	Phosphorus Trichloride
Chloroacetaldehyde	Potassium Chlorite
Chloroacetic Acid	Potassium Fluoride
Chromic Acid	Potassium Fluosilicate
Dimethyl Sufate	Propionaldehyde
Fluorine	Resorcinol
Fluosilicic Acid	Sodium Chlorite
Formic Acid	Sodium Fluoride
Hydriodic Acid	Sodium Fluosilicate
Hydrochloric Acid	Sulfur Dioxide
Hydrofluoric Acid	Sulfur Trioxide
Hydrogen Bromide	Sulfuric Acid
Hydrogen Chloride	Sulfurous Acid
Hydrogen Peroxide	Trichloroacetic Acid
Isobutyraldehyde	

Your Goal is: To dilute the acid in the stomach and prevent further injury caused by vomiting.

1. Remove the victim from the contaminated area to a quiet, well ventilated area.

2. Have someone call a poison control center, inform them of the name of the chemical swallowed and follow their advice.

3. Have someone call the Emergency Medical Service and arrange for transport to a medical facility.

A. If the victim stops breathing:

1. Wipe the mouth and rinse away any remaining chemical. Check the airway for obstruction. Provide ventilation using a bag-valve mask or the chest pressure-arm lift technique.

B. If the victim's face is BLUE or if respiration is labored:

1. Check the airway for obstruction.

2. Give him oxygen to breath by mask if available.

C. If the victim is unconscious:

1. Lay him on his left side and loosen his collar and belt.

D. If the victim is conscious:

1. Loosen tight clothing around the neck and waist.

2. Have him rinse his mouth several times with cold water and spit out.

3. You may give him one or two cups of water or milk. You may also give gastric antacids such as milk of magnesia or aluminum hydroxide. Follow the poison control center's directions if they differ. Stop if the victim becomes nauseated.

4. Keep the victim warm and quiet.

DO NOT give an unconscious or a convulsing person anything to drink.

DO NOT induce vomiting

DO NOT give the victim any oils.

DO NOT try to neutralize the acid with a strong base.

DO NOT give sodium bicarbonate or any carbonated drinks.

First Aid in Case of Ingestion of:

Acetaldehyde
Acrolein
Acrylamide
Aldicarb
Alkali Dichromates
Alkali Meta-Borates
Allyl Alcohol
Allyl Chloride
Allyl Glycidyl Ether
Allyl Propyl Disulfide
Aluminum Chloride
Aluminum Trichloride
Antimony and Compounds
Benzyl Chloride
Bis (Chloromethyl) Ether
Boric Acid
Butyl Glycidyl Ether (n-)
Cadmium (dust and fumes) (metal)
Calcium Dichromate
Caprolactam
Carbamates
Carbon Disulfide
Cement
Chlorinated Lime
Chloro-1-Nitropropane (1-)
Chloroacetophenone (2-)
Chlorobenzylidene Malonitrile
Chloropicrin
Chlorthion
Chromium Chloride
Copper Chloride
Copper Sulfate
Creosote
Cresols
Crotonaldehyde
Cyclohexanol
Cyclohexanone
DDVP
Demeton
Diazinon
Diazomethane
Dibutyl Phthalate
Dichloro-5,5-Dimethylhydantoin
Diepoxybutane

Diethylaminoethanol
Diglycidyl Ether
Dipterex
Dipyridyl Chloride
Dipyridyl Dimethyl Sulfate
Diquat
EPN
Epichlorohydrin
Ethyl Acrylate
Ethyl Chloroformate
Ethylene Oxide
Ethyleneimine
Formaldehyde
Furfural
Glutaraldehyde
Glycidol
Glycidyl Acrylate
Gramoxone
Iodine
Iron Chloride
Isopestox
Lead (dust and fumes)
Lead Acetate
Lead Antimonate
Lead Arsenate
Lead Carbonate
Lead Chromate
Lead Chromate (yellow)
Lead Dioxide
Lead Nitrate
Lead Oxide (PbO)
Lead Oxide (red)
Lead Oxychloride
Lead Subacetate
Lead Sulfide
Leptophos
Malathion
Maleic Anhydride
Mercuric Chloride
Mercuric Iodide (red)
Mercurous Chloride
Mercurous Iodide
Mercury (metal)
Mercury (soluble salts)

Mercury Acetate
Mercury Nitrate
Mercury Oxycyanide
Methyl Acrylate
Methyl Chloroformate
Methyl Isocyanate
Methyl Methacrylate Monomer
Methyl Parathion
Methylenebis (Phenyl Isocyanate)
Nickel (fumes and dust)
Nickel Carbonyl
OMPA
Organophosphate Compounds
Paraoxon
Parathion
Perborates
Perchloromethyl Mercaptan
Phenylenediamine (p-)
Phorate
Phosdrin
Phosphoric Ester
Phthalic Anhydride
Picric Acid
Platinum and Compounds
Potassium Chromate

Potassium Dichromate
Propylene Glycol Monomethyl Ether
Propylene Oxide
Pyrethrins
Quaternary Ammonium Compounds
Quinone
Ronnel
Sodium Borate
Sodium Chromate
Sodium Dichromate
Sodium Hypochlorite
Styrene
Sulfotepp
TEPP
Tetryl
Titanium Chlorides
Toluene 2,4-di-Isocyanate
Toluene 2,6-di-Isocyanate
Tributyl Phosphate
Trimellitic Anhydride
Trithion
Uranium and Compounds
Vanadium and Compounds
Zinc Chloride

Your Goal is: To empty the stomach and prevent further injury caused by absorption.

1. Remove the victim from the contaminated area to a quiet, well ventilated area.

2. Have someone call a poison control center, inform them of the chemical swallowed and follow their advice.

3. Have someone call the Emergency Medical Service and arrange for transport to a medical facility.

A. If the victim stops breathing:

1. Administer mouth to mouth respiration, being sure to wipe away any remaining chemical.

B. If the victim's face is BLUE or if respiration is labored:

1. Check the airway for obstruction.

2. Give the victim oxygen to breath by mask if available.

C. If the victim is unconscious:

1. Lay him on his left side and loosen his collar and belt.

E. If the victim is conscious:

1. Loosen tight clothing around the neck and waist.

2. Have him rinse his mouth several times with cold water and spit out.

3. Give him one or two cups of water or milk to drink.

4. Unless advised otherwise by the poison control center, induce vomiting by giving 2 tablespoons of syrup of ipecac (adult dose) followed by a cup of water.

5. If vomiting does not occur after 10 minutes or if you do not have syrup of ipecac, induce vomiting by asking him to touch the back of his throat with his finger or with a spoon handle or blunt object.

6. Have the victim sit up and lean forward while vomiting.

7. Save vomitus for analysis later.

8. AFTER the victim vomits, give the victim a mixture of 2 tablespoonfuls of activated charcoal mixed in 8 oz. of water to drink.

DO NOT give an unconscious person anything to drink.

DO NOT give activated charcoal before or with syrup of ipecac.

DO NOT give the victim any oils.

DO NOT give alcohol, drugs or stimulants like tea or coffee.

DO NOT give sodium bicarbonate or carbonated drinks.

3

First Aid in Case of Ingestion of:

Aliphatic Amines	Lithium Hydride
Ammonia	Methylamine
Ammonium Carbonate	Milk of Lime
Ammonium Hydroxide	Potassium
Ammonium Sulfide	Potassium Carbonate
Butylamine	Potassium Hydroxide
Calcium Carbide	Potassium Oxide
Calcium Hydroxide	Propylamine
Calcium Oxide	Sodium
Dibutylamine	Sodium Carbonate
Diethylamine	Sodium Hydroxide
Dimethylamine	Sodium Oxide
Dipropylamine	Sodium Peroxide
Ethanolamine	Sodium Silicate
Ethylamine	Triethylamine
Isopropylamine	Trimethylamine
Lime	Trisodium Phosphate
Lithium Carbonate	

Your Goal is: To dilute the chemical in the stomach and prevent further injury caused by vomiting.

1. Remove the victim from the contaminated area to a quiet, well ventilated area.

2. Ask someone to call the Emergency Medical Service and arrange for transport.

3. Call a poison control center, inform them of the chemical swallowed and follow their advice.

A. If the victim stops breathing:

1. Check the airway for obstruction. Wipe away any remaining chemical in the mouth.

2. Use a bag-valve mask or the chest pressure-arm lift technique to provide artificial respiration.

B. If the victim's face is BLUE or if respiration is labored:

1. Check the airway for obstruction.

2. Give him oxygen to breath by mask if available.

C. If the victim is unconscious:

1. Lay him on his left side and loosen his collar and belt.

2. Do not leave him unattended.

D. If the victim is conscious:

1. Loosen tight clothing around the neck and waist.

2. Have him rinse his mouth several times with cold water and spit out.

3. Give him 1 or 2 cups of milk. Stop if the victim becomes nauseated.

4. Keep the victim warm and quiet.

DO NOT induce vomiting.

DO NOT give an unconscious person or a convulsing person anything to drink.

DO NOT give the victim any oils.

DO NOT try to neutralize a base with an acid.

DO NOT give sodium bicarbonate or carbonated drinks.

First Aid in Case of Ingestion of:

Benzene
Cumene
Cyclohexane
Decane
Ethyl Ether
Gasoline
Heptane
Hexane
Liquefied Petroleum Gas
Naptha

Nitromethane
Nonane
Octane
Pentane
Petroleum Ethers
Stoddard Solvent
Toluene
Trifluoroethane
Turpentine
Xylene

Your Goal is: To trap the chemical in the stomach and prevent further injury caused by vomiting.

1. Remove the victim from the contaminated area to a quiet, well ventilated area, away from any fire or smoke.

2. Call a poison control center, inform them of the chemical swallowed and follow their advice.

3. Call the Emergency Medical Service and arrange for transport.

A. If the victim stops breathing:

1. Wipe away any remaining material off the lips.

2. Clear the airway and administer mouth to mouth respiration. Avoid inhaling the exhaled air of the victim.

B. If the victim's face is BLUE or if respiration is labored:

1. Check the airway for obstruction.

2. Give him oxygen to breath by mask if available.

C. If the victim has a seizure or convulsion:

1. Do not attempt to restrain him but position him in such a way that he will not injure himself.

2. Watch for airway obstruction and try to reposition the head if it occurs.

3. After the convulsion, place the victim on his side.

D. If the victim is unconscious:

1. Lay him on his left side and loosen his collar and belt.

E. If the victim is conscious:

1. Loosen tight clothing around the neck and waist.
2. Have him rinse his mouth several times with cold water and spit out.
3. Give him a mixture of 2 tablespoons of activated charcoal mixed in 8 oz. of water to drink.
4. Keep the victim warm and quiet.

DO NOT give an unconscious person or a convulsing person anything to drink.

DO NOT induce vomiting.

DO NOT give the victim any oils.

DO NOT give the victim any alcohol, drugs or stimulants like coffee or tea.

DO NOT give sodium bicarbonate or carbonated drinks.

DO NOT force a hard object between the victim's teeth during a convulsion.

5

First Aid in Case of Ingestion of:

Aldrin
Aminopyridine
Bromoform
Butyltoluene
Camphor
Carbon Tetrachloride
Chlordane
Chlorobenzene
Chlorobromomethane
Chloroethane
Chloroform
Chloropropane
Chloropropene
Decaborane
Dichlorobenzene
Dichloroethane
Dichloroethylene
Dichloropropane
Dichlorotetrafluoroethane
Dimethylhydrazine (1,1-)

Ethylene Dichloride
Freon 112, 113, 114, 115
Halothane
Hexachloroethane
Hydrazine
Lindane
Methylchloroform
Methylene Chloride
Organochlorines
Paraquat
Pentaborane
Pentachloroethane
Tetrachlorodifluoroethane
Tetrachloroethane
Tetrachloroethylene
Trichloroethane
Trichloroethylene
Trichlorotrifluoroethane
Vinylidene Chloride

Your Goal is: To empty the stomach and prevent further injury caused by absorption.

1. Remove the victim from the contaminated area to a quiet, well ventilated area.

2. Have someone call a poison control center, inform them of the chemical swallowed and follow their advice.

3. Monitor the victim's heartbeat by taking his pulse every few minutes. If the heartbeat is irregular or very slow, be prepared to administer CPR.

4. Have someone call the Emergency Medical Service and arrange for transport to a medical facility.

A. If the victim stops breathing:

1. Wipe away any remaining material from the mouth and face and avoid inhaling the exhaled air of the victim.

2. Clear the airway and administer mouth to mouth respiration. If this is not possible, use a bag-valve mask or the chest pressure-arm lift technique.

B. If the victim's face is BLUE or if respiration is labored or shallow:

1. Check the airway for obstruction.

2. Give the victim oxygen to breath by mask if available.

C. If the victim is unconscious:

1. Lay him on his left side and loosen his collar and belt.

D. If the victim is conscious:

1. Loosen tight clothing around the neck and waist.

2. Keep the victim quiet and calm.

3. Unless advised otherwise by the poison control center, induce vomiting by giving 2 tablespoons of syrup of ipecac (adult dose) followed by a cup of water.

4. If you do not have syrup of ipecac or if vomiting does not occur after 10 minutes, induce vomiting by asking him to touch the back of his throat with his finger or with a spoon handle or blunt object.

5. Have the victim sit up and lean forward while vomiting.

6. Save vomitus for analysis later.

7. **AFTER** the victim vomits, give him a mixture of 2 tablespoonfuls of activated charcoal mixed in 8 oz. of water to drink.

DO NOT give any stimulants like tea or coffee.

DO NOT administer epinephrine or any medications for asthma to someone who has ingested these chemicals.

DO NOT give activated charcoal before or with syrup of ipecac.

First Aid in Case of Ingestion of:

Barium (soluble salts)   Barium Hydroxide
Barium Acetate   Barium Nitrate
Barium Carbonate   Barium Oxide
Barium Chloride   Barium Sulfide

Your Goal is: To empty the stomach and prevent further injury caused by absorption.

1. Remove the victim from the contaminated area to a quiet, well ventilated area.

2. Call a poison control center, inform them of the chemical swallowed and follow their advice.

3. Monitor the victim's heartbeat by taking his pulse every few minutes. If the heartbeat is irregular, be prepared to administer CPR.

4. Call the Emergency Medical Service and arrange for transport to a medical facility.

A. If the victim stops breathing:

1. Clear the airway and administer mouth to mouth respiration being sure to wipe and rinse away any remaining chemical.

2. Give the victim oxygen to breathe, by mask, if available.

B. If the victim's face is BLUE or if respiration is labored or shallow:

1. Check the airway for obstruction.

2. Give the victim oxygen to breath by mask if available.

C. If the victim is unconscious:

1. Lay him on his left side and loosen his collar and belt.

D. If the victim is conscious:

1. Loosen tight clothing around the neck and waist.

2. Keep the victim quiet and calm.

3. Unless advised otherwise by the poison control center, induce vomiting by giving 2 tablespoons of syrup of ipecac (adult dose) followed by a cup of water.

4. If you do not have syrup of ipecac or if vomiting does not occur after 10 minutes, induce vomiting by asking him to touch the back of his throat with his finger or with a spoon handle or blunt object.

5. Have the victim sit up and lean forward while vomiting.

6. Save vomitus for analysis later.

7. **AFTER** the victim vomits, give him a mixture of 1 tablespoonful of magnesium sulfate mixed in 8 oz. of water to drink.

DO NOT leave the victim alone.

DO NOT give an unconscious person or a person who is convulsing anything to drink.

7

First Aid in Case of Ingestion of:

Aniline	Nitrobenzene
Anisidines (ortho)	Nitrochlorobenzene (p-)
Anisidines (para)	Nitroglycerin
Benzidine	Nitrocresolic Herbicides
Chlorophenoxy Compounds	Nitrophenolic Herbicides
DNBP	Nitrophenols
DNOC	Nitrotoluene
Dimethylaniline	Pentachlorophenate
Dinitrobenzene	Pentachlorophenol
Dinitrocresols	Phenylhydrazine
Dinitrophenols	Phenylhydroxylamine
Dinitrotoluene	Phenylnaphthylamine
Ethyl Nitrate	Propyl Nitrate
Ethylene Glycol Dinitrate	Tetranitromethane
Hydroquinone	Tolidine (o-)
Methyl Nitrate	Toluidine
Monomethylhydrazine	Trinitrobenzene
Napthalene	Trinitrotoluene
Naphthylamines	Xylidine
Nitroanilines	

Your Goal is: To empty the stomach and prevent further injury caused by absorption.

1. Remove the victim from the contaminated area to a quiet, well ventilated area.

2. Call a poison control center, inform them of the chemical swallowed and follow their advice.

3. Call the Emergency Medical Service and arrange for transport to a medical facility.

A. If the victim stops breathing:

1. Wipe away any remaining chemical from the lips.

2. Clear the airway and administer mouth to mouth respiration.

B. If the victim's face is BLUE or BROWNISH or if respiration is labored:

1. Check the airway for obstruction.

2. Give the victim oxygen to breath by mask if available.

C. If the victim is unconscious:

1. Lay him on his left side and loosen his collar and belt.
2. Do not leave him unattended.

D. If the victim is conscious:

1. Loosen tight clothing around the neck and waist.
2. Have him rinse his mouth several times with cold water and spit out.
3. Give him 1 or 2 cups of water to drink.
4. Unless advised otherwise by the poison control center, induce vomiting by giving 2 tablespoons of syrup of ipecac (adult dose) followed by a cup of water.
5. If you do not have syrup of ipecac or if vomiting doesn't occur after 10 minutes, induce vomiting by asking him to touch the back of his throat with his finger or with a spoon handle or a blunt object.
6. Have the victim sit up and lean forward while vomiting.
7. Save vomitus for analysis later.
8. **AFTER** the victim vomits, give the victim a mixture of 2 tablespoonfuls of activated charcoal mixed in 8 oz. of water to drink. Give as much water as he wants.
9. Give oxygen to breath by mask if available.

E. If the victim is feverish:

1. Bathe the hands and head in cool water or wrap the legs with towels soaked in water.

DO NOT give an unconscious person anything to drink.

DO NOT give activated charcoal before or with syrup of ipecac.

DO NOT give the victim any oils, milk, eggs or alcohol.

DO NOT give drugs like aspirin.

DO NOT give sodium bicarbonate or carbonated drinks.

8

First Aid in Case of Ingestion of:

Ammonium Chlorate	Potassium Perchlorate
Ammonium Perchlorate	Sodium Bicarbonate
Calcium Chloride	Sodium Chlorate
Dithiocarbamates	Sodium Chloride
Potassium Chlorate	Sodium Perchlorate
Potassium Chloride	Thiocarbamates

Your Goal is: To empty the stomach and prevent further injury caused by absorption.

1. Remove the victim from the contaminated area to a quiet, well ventilated area.

2. Call a poison control center, inform them of the chemical swallowed and follow their advice.

3. Loosen tight clothing around the neck and waist.

4. Have him rinse his mouth several times with cold water and spit out.

5. Give him 1 or 2 cups of water or milk to drink.

6. Unless advised otherwise by the poison control center, induce vomiting by giving 2 tablespoons of syrup of ipecac (adult dose) followed by a cup of water.

7. If you do not have syrup of ipecac or if vomiting does not occur in 10 minutes, induce vomiting by asking him to touch the back of his throat with his finger or with a spoon handle or blunt object.

8. Have the victim sit up and lean forward while vomiting.

9. Save vomitus for analysis later.

10. **AFTER** the victim vomits, give the victim a mixture of 2 tablespoonfuls of activated charcoal mixed in 8 oz. of water to drink. Give as much water as he wants after this.

11. Call the Emergency Medical Service and arrange for transport to a medical facility.*

DO NOT give activated charcoal before or with syrup of ipecac.

*Reminder: In case of ingestion of perchlorates, the physician will keep the patient under observation for several days, giving special attention to kidney function.

DO NOT give the victim any oils.

DO NOT give alcohol, drugs or stimulants like tea or coffee.

DO NOT give an unconscious person anything to drink.

9

First Aid in Case of Ingestion of:

Dibutyllead
Dibutyltin
Diethyllead
Diethylmercury
Diethyltin
Dihexyltin
Diiododiethyltin
Dimethylmercury
Dimethyltin
Dioctylin
Ethylmercuric Chloride
Ethylmercuric Hydroxide
Ethylmercury
Lead Oleate
Lead Phenate
Lead Phthalate
Lead Stearate
Mercury (organic compounds)
Mercury Fulminate
Methylmercury
Methylmercury Borate

Methylmercury Hydroxide
Methylmercury Iodide
Methylmercury Nitrate
Methylmercury Phosphate
Phenylmercuric Acetate
Phenylmercury
Phenylmercury Oleate
Tetrabutyltin
Tetraethyllead
Tetraethyltin
Tetraisoalkyltin
Tetramethyllead
Tetrapentyltin
Tetrapropyltin
Tributyllead
Tributyltin
Triethyllead
Trimethyllead
Trimethyltin
Triphenyltin
Tripropyltin

Your Goal is: To empty the stomach and prevent further injury caused by absorption.

1. Remove the victim from the contaminated area to a quiet, well ventilated area.

2. Call a poison control center, inform them of the chemical swallowed and follow their advice.

3. Call the Emergency Medical Service and arrange for transport to a medical facility.

4. Loosen tight clothing around the neck and waist.

5. Have the victim rinse his mouth several times with cold water and spit out.

6. Give him 1 or 2 cups of water or milk to drink.

7. Unless advised otherwise by the poison control center, induce vomiting by giving 2 tablespoons of syrup of ipecac (adult dose) followed by a cup of water.

8. If you do not have syrup of ipecac or if vomiting does not occur in 10 minutes, induce vomiting by asking him to touch the back of his throat with his finger or with a spoon handle or blunt object.

9. Have the victim sit up and lean forward while vomiting.

10. Save vomitus for analysis later.

11. **AFTER** the victim vomits, give him a mixture of 2 tablespoonfuls of activated charcoal mixed in 8 oz. of water to drink.

12. Do not leave the victim unattended.

DO NOT give an unconscious person anything to drink.

DO NOT give activated charcoal before or with syrup of ipecac.

DO NOT give the victim any oils.

DO NOT give alcohol, drugs or stimulants like tea or coffee.

10

First Aid in Case of Ingestion of:

Alkanes (liquids/solids)
Aluminum (dust)
Aluminum Hydrate
Aluminum Hydroxide
Aluminum Oxide
Asbestos
Calcium Carbonate
Carbon
Carbon Black
Chloronaphthalenes
Diphenyl
Diphenylamine
Fibrous Glass

Hexachlorobenzene
Kaolin
Paraffins
Polybrominated Biphenyls (PBBs)
Polychlorinated Biphenyls (PCBs)
Polyvinyl Chloride
Silica
Talc
Titanium (dust and fumes)
Titanium Dioxide
Tungsten Carbide
Yttrium and Compounds

Your Goal is: To empty the stomach and prevent further injury caused by absorption.

1. Remove the victim from the contaminated area to a quiet, well ventilated area.

2. Call a poison control center, inform them of the chemical swallowed and follow their advice.

3. Call the Emergency Medical Service and arrange for transport to a medical facility.

4. Loosen tight clothing around the neck and waist.

5. Have the victim rinse his mouth several times with cold water and spit out.

6. Give him 1 or 2 cups of water or milk to drink.

7. Unless advised otherwise by the poison control center, induce vomiting by giving 2 tablespoons of syrup of ipecac (adult dose) followed by a cup of water.

8. If you do not have syrup of ipecac or if vomiting does not occur in 10 minutes, induce vomiting by asking him to touch the back of his throat with his finger or with a spoon handle or blunt object.

9. Have the victim sit up and lean forward while vomiting.

10. Save vomitus for analysis later.

11

First Aid in Case of Ingestion of:

Acetone Cyanohydrin	Ferrocyanides
Acetonitrile	Hydrocyanic Acid
Acrylonitrile	Isobutyronitrile
Adiponitrile	Malononitrile
Bitter Almond Oil (Amygdalin)	Methacrylonitrile
Cherry Laurel Water	Nitroferricyanides (salts)
Cyanogen Bromide	Potassium Cyanide
Cyanogen Chloride	Sodium Cyanide
Cyanogen Iodide	Tetramethyl Succinonitrile
Ferricyanides	

Your Goal is: To start lifesaving treatment, call for help and, if possible, empty the stomach and prevent further injury caused by absorption. PROMPT TREATMENT IS LIFESAVING.

1. Ask someone to call a poison control center, inform them of the chemical swallowed and follow their advice.

2. Ask someone to call the Emergency Medical Service and arrange for transport to a medical facility.

A. If the victim is unconscious or unresponsive:

1. Break perles of amyl nitrite in a handkerchief one at a time every 5 minutes and hold under the victim's nose for 30 seconds and remove for 30 seconds. Break a fresh perle every 5 minutes until 3 or 4 perles are used up.

2. Lay him on his left side and loosen his collar and belt.

3. Check the airway for obstruction.

4. Give the victim oxygen to breathe by mask if available.

B. If the victim stops breathing:

1. Administer artificial respiration using a bag-valve mask or the chest pressure-arm lift technique.

2. Place a handkerchief with a broken amyl nitrite perle in it inside the mask or over the victim's face while performing artificial respiration. See A, 1.

C. If the victim is conscious and alert:

1. Break perles of amyl nitrite in a handkerchief one at a time every 5 minutes and hold under the victim's nose for 30 seconds, then remove for 30 seconds. Use no more than 4 perles.

2. Remove him from the contaminated area to a quiet, well ventilated area.

3. Loosen tight clothing around the neck and waist.

4. Have him rinse his mouth several times with cold water and spit out.

5. Give him 1 or 2 cups of water or milk to drink.

6. Induce vomiting by touching the back of the throat with your finger, a spoon handle or a blunt object.

7. Have the victim sit up and lean forward while vomiting.

8. Save vomitus for analysis later. Avoid skin contact with it.

9. Do not leave the victim alone.

DO NOT give an unconscious person or a person who is having a convulsion anything to drink.

DO NOT give alcohol, drugs or stimulants like tea or coffee.

DO NOT continue to try to induce vomiting in someone who doesn't gag when you touch the back of his throat.

12

First Aid in Case of Ingestion of:

Acetone	Glycerin
Aliphatic Alcohols—Amyl	Heptanol
Aliphatic Alcohols—Butyl	Hexanol
Amyl Acetate	Isobutyl Acetate
Butanol	Isopropyl Acetate
Butyl Acetate	Isopropyl Alcohol
Decanol	Methyl Acetate
Diacetone Alcohol	Methyl Alcohol
Diethylene Glycol	Methyl n-Butyl Ketone
Diisobutylcarbinol	Methyl Ethyl Ketone
Dioxane	Methyl Isoamyl Ketone
Ethyl Acetate	Methyl Isobutyl Ketone
Ethyl Alcohol	Methyl Isopropyl Ketone
Ethylbenzene	Pentanol
Ethylene Chlorohydrin	Propyl Acetate
Ethylene Glycol	Propyl Alcohol
Ethylene Glycol Monomethyl Ether	Propylene Glycol
Ethylhexyl Acetate	Triethylene Glycol
Furfuryl Alcohol	Vinyl Acetate

Your Goal is: To empty the stomach and prevent further injury caused by absorption.

1. Remove the victim from the contaminated area to a quiet, well ventilated area.

2. Call a poison control center, inform them of the chemical swallowed and follow their advice.

3. Call the Emergency Medical Service and arrange for transport to a medical facility.

A. If the victim stops breathing:

1. Administer mouth to mouth respiration being sure to wipe and rinse away any remaining chemical. If this is not possible, use a bag-valve mask or the chest pressure-arm lift technique.

2. Give him oxygen to breathe by mask, if available.

B. If the victim's face is BLUE or if respiration is labored or shallow:

1. Check the airway for obstruction.

2. Give him oxygen to breathe by mask if available.

C. If the victim is unconscious:

1. Lay him on his left side and loosen his collar and belt.

D. If the victim is conscious:

1. Loosen tight clothing around the neck and waist.

2. Keep the victim quiet and calm.

3. Unless advised otherwise by the poison control center, induce vomiting by giving 2 tablespoons of syrup of ipecac (adult dose) followed by a cup of water.

4. If you do not have syrup of ipecac or if vomiting doesn't occur in 10 minutes, induce vomiting by asking the victim to touch the back of his throat with his finger or with a spoon handle or blunt object.

5. Have the victim sit up and lean forward while vomiting.

6. Save vomitus for analysis later.

DO NOT try to give an unconscious person anything to drink.

DO NOT give any stimulants like tea or coffee.

First Aid in Case of Skin Contact with:

Acetaldehyde
Acetic Acid
Acetic Anhydride
Acrolein
Ammonium Sulfide
Antimony and Compounds
Arsenic Trichloride
Arsenicals
Arsine
Benzyl Chloride
Boron Trifluoride
Bromine
Butyraldehyde
Carbon Disulfide
Chlorine
Chlorine Dioxide
Chloroacetaldehyde
Chloroacetic Acid
Creosote
Cresols
Crotonaldehyde
Dichloro-5,5-Dimethylhydantoin
Dimethyl Sulfate
Ethyl Chloroformate
Formaldehyde
Formic Acid
Hydriodic Acid
Hydrochloric Acid
Hydrogen Bromide
Hydrogen Chloride
Hydrogen Peroxide
Hydrogen Selenide
Hydrogen Sulfide
Iodine
Isobutyraldehyde
Maleic Anhydride

Methyl Chloroformate
Methyl Mercaptan
Nitric Acid
Nitric Oxide
Nitrogen Dioxide
Osmic Acid
Oxalic Acid
Ozone
Peracetic Acid
Perchloric Acid
Phenol
Phenylenediamine (p-)
Phosgene
Phosphine
Phosphoric Acid
Phosphorus
Phosphorus Chlorides
Phosphorus Pentachloride
Phosphorus Pentasulfide
Phosphorus Trichloride
Potassium Chlorite
Propionaldehyde
Quinone
Resorcinol
Selenium Hexafluoride
Silane
Sodium Chlorite
Stibine
Sulfur Dioxide
Sulfur Trioxide
Sulfuric Acid
Sulfurous Acid
Tellurium Hexafluoride
Tributyl Phosphate
Trichloroacetic Acid

Your Goal is: To remove all chemical from contact with the victim's skin as quickly as possible. A delay of only seconds may increase the injury.

1. Remove the victim from the source of contamination and take him **IMMEDIATELY** to the nearest shower or source of clean water.

2. Remove clothing, shoes, socks and jewelry from the affected areas as quickly as possible, cutting them off if necessary. **BE CAREFUL** not to get any of the chemical on your skin or clothing. **BE CAREFUL** not to inhale emitted vapors.

3. Blot excess chemical from the skin very gently and without delay.

4. In the case of extensive splashing of the product, wash the victim down under a cold or luke-warm shower or a hand-held hose, at the same time protecting his eyes.

5. Wash the affected area under tepid running water using a mild soap.

6. Rinse the affected area with tepid water for at least 15 minutes.

7. Dry the skin very gently with a clean, soft towel.

In case of burns (inflammation, blisters, or lesions) and in the absence of medical personnel:

8. Notify a physician, emergency room, or poison control center and inform them of the nature of the substance and the accident.

9. Loosely apply a dry sterile dressing if available or use a clean dry cloth.

10. Dress the victim in clean clothes or cover him with a sheet.

11. Elevate the affected area above the level of the victim's heart.

12. Arrange for transport to the nearest medical facility.

If the victim is in pain:

13. Immerse painful area in cold water or apply cold wet dressings on the burned area.

If the victim is in a state of shock:

14. Lay him down on his side and cover him with a blanket.

15. Elevate his feet.

DO NOT break open blisters or remove skin. If clothing is stuck to the skin after flushing with water, do not remove it.

DO NOT rub or apply pressure to the affected skin.

DO NOT apply any oily substance to the affected skin.

DO NOT use hot water.

Aldrin
Alkali Dichromates
Alkali Meta-Borates
Aluminum Chloride
Aluminum Trichloride
Aminopyridine
Ammonium Chlorate
Ammonium Perchlorate
Arsenic
Barium (soluble salts)
Barium Acetate
Barium Carbonate
Barium Chloride
Barium Hydroxide
Barium Nitrate
Barium Oxide
Barium Sulfide
Boric Acid
Cadmium (dust and fumes) (metal)
Calcium Chloride
Calcium Dichromate
Calcium Hypochlorite
Camphor
Caprolactam
Chlordane
Chlorinated Lime
Chromic Acid
Chromium Chloride
Copper Chloride
Copper Sulfate
Decaborane
Dibutyltin
Diethylmercury
Diethyltin
Dihexyltin
Diiododiethyltin
Dimethylhydrazine (1,1-)
Dimethylmercury
Dimethyltin
Dioctyltin
Dithiocarbamates
Ethylmercuric Chloride
Ethylmercuric Hydroxide
Ethylmercury

Iron Chloride
Lead (dust and fumes)
Lead Acetate
Lead Antimonate
Lead Arsenate
Lead Carbonate
Lead Chromate
Lead Chromate (yellow)
Lead Dioxide
Lead Nitrate
Lead Oxide (PbO)
Lead Oxide (red)
Lead Oxychloride
Lead Subacetate
Lead Sulfide
Lindane
Mercuric Chloride
Mercuric Iodide (red)
Mercurous Chloride
Mercurous Iodide
Mercury (metal)
Mercury (organic compounds)
Mercury (soluble salts)
Mercury Acetate
Mercury Fulminate
Mercury Nitrate
Mercury Oxycyanide
Methylmercury
Methylmercury Borate
Methylmercury Hydroxide
Methylmercury Iodide
Methylmercury Nitrate
Methylmercury Phosphate
Organochlorines
Pentaborane
Perborates
Phenylmercuric Acetate
Phenylmercury
Phenylmercury Oleate
Phthalic Anhydride
Platinum and Compounds
Potassium Chlorate
Potassium Chloride
Potassium Chromate

Potassium Dichromate
Potassium Perchlorate
Sodium Bicarbonate
Sodium Borate
Sodium Chlorate
Sodium Chloride
Sodium Chromate
Sodium Dichromate
Sodium Hypochlorite
Sodium Perchlorate
Tetrabutyltin
Tetraethyltin
Tetraisoalkyltin

Tetrapentyltin
Tetrapropyltin
Thiocarbamates
Titanium Chlorides
Tributyltin
Trimellitic Anhydride
Trimethyltin
Triphenyltin
Tripropyltin
Uranium and Compounds
Vanadium and Compounds
Zinc Chloride

Your Goal is: To remove all chemical from contact with the victim's skin.

1. Remove the victim from the source of contamination.

2. Remove clothing, shoes, socks and jewelry from affected areas.

3. Wash the affected area under tepid running water.

4. Rinse carefully until no traces of the chemical can be seen.

5. Dry gently with a clean, soft towel.

6. Notify a physician, emergency room, or poison control center of the accident and inform them of the nature of the substances.

If the skin is red, swollen or painful:

7. Dress the victim in clean clothes or cover him with a sheet.

8. Immerse painful area in cold water or apply cold, wet dressings to the burned area.

9. Arrange to transport the victim to the hospital as soon as possible.

DO NOT break open blisters or remove skin. If clothing is stuck to the skin after flushing with water, do not remove it.

DO NOT rub or apply pressure to the affected skin.

DO NOT apply any oily substance to the affected skin.

DO NOT use hot water.

3

First Aid in Case of Skin Contact with:

Barium Fluoride
Chlorine Trifluoride
Fluorine
Fluosilicic Acid
Hydrofluoric Acid
Nitrogen Trifluoride

Oxygen Difluoride
Perchloryl Fluoride
Potassium Fluoride
Potassium Fluosilicate
Sodium Fluoride
Sodium Fluosilicate

Your Goal is: To remove all chemical from contact with the victim's skin as quickly as possible. A delay of only seconds may increase the injury.

1. Remove the victim from the source of contamination and take him **IMMEDIATELY** to the nearest shower or source of clean water.

2. Protect yourself by wearing rubber gloves and air-tight safety goggles.

3. Remove the clothing from affected areas under the shower (clothing can be cut away, if necessary). Care should be taken not to contaminate healthy skin or eyes. If the victim is already wearing air-tight safety goggles, do not remove them.

4. Wash him down with cold water for 15 minutes or longer.

5. Pat the skin dry very gently with a clean, soft towel.

6. Apply calcium gluconate gel 2.5%, if available, on the affected skin.

7. Massage the gel gently into every burnt area with clean gloved fingers.

8. Dress the victim in clean clothes or cover with a sheet.

9. Notify a physician, emergency room, or poison control center and inform them of the nature of the substance and the accident.

In the case of burns (inflammation, blisters, or painful lesions) and in the absence of medical personnel:

10. If there is no calcium gluconate gel available, rub a clean cube of ice on the painful areas and apply a dry, sterile dressing if available or use a clean dry cloth.

11. Call the Emergency Medical Service and arrange for transport to the nearest medical facility.

If the victim shows signs of shock:

12. Cover him with a blanket.

13. Lay him down in a quiet place on his side with legs raised.

NOTE: Oral administration of 6 tablets of effervescent calcium gluconate dissolved in water is recommended in case of large area burns. Give only at direction of a physician.

4

First Aid in Case of Skin Contact with:

Aliphatic Amines	Isopropylamine
Ammonia	Lime
Ammonium Hydroxide	Methylamine
Butylamine	Potassium
Calcium Carbide	Potassium Hydroxide
Calcium Oxide	Potassium Oxide
Cement	Propylamine
Dibutylamine	Sodium
Diethylamine	Sodium Hydroxide
Dimethylamine	Sodium Oxide
Dipropylamine	Sodium Peroxide
Ethanolamine	Triethylamine
Ethylamine	Trimethylamine

Your Goal is: To remove all chemical from contact with the victim's skin as quickly as possible. A delay of only seconds may increase the injury.

1. Remove the victim from the source of contamination and take him **IMMEDIATELY** to the nearest shower or source of clean water.

2. While wearing polyvinyl gloves and air tight safety goggles, remove the victim's clothing, shoes, socks and jewelry from the affected areas as quickly as possible, cutting them off if necessary. Do this under the shower or while flushing with water. Be careful not to get any of the chemical on your skin or clothing. Take care not to contaminate the healthy skin and eyes of the victim. If the victim is wearing air tight safety goggles, don't remove them.

3. Wash the victim down under a cold or luke-warm shower or a hand held hose, at the same time protecting his eyes.

4. Continue to wash the affected area under luke-warm running water until the feeling of stickiness or soapiness caused by the caustic chemical disappears. This may take an hour or more.

5. Dry the skin very gently with a clean, soft towel.

6. Notify a physician, emergency room, or poison control center and inform them of the nature of the substance and the accident.

In case of inflammation, blisters, breaks in the skin or pain:

7. Loosely apply a dry sterile dressing if available or use a clean dry cloth.

8. Dress the victim in clean clothes or cover him with a sheet.

9. Call the Emergency Medical Service and arrange for transport to the nearest medical facility.

If the victim is in a state of shock;

10. Lay him on his side and cover him with a blanket.

11. Elevate his feet.

DO NOT break open blisters or remove skin. If clothing is stuck to the skin after flushing with water, do not remove it.

DO NOT rub or apply pressure to the affected skin.

Acetylene
Alkanes (gasses, C_1 to C_4)
Butadiene
Butane
Carbon Dioxide
Carbon Dioxide Snow
Chlorodifluoroethane
Chlorodifluoromethane
Chloroethane
Chlorofluoroethane
Chlorofluoromethane
Chloromethane
Chlorotrifluoroethylene
Chlorotrifluoromethane
Dichlorodifluoromethane
Dichloroethane
Dichlorofluoromethane
Difluoroethanes
Difluoroethylene
Ethane
Ethyl Ether
Ethyl Fluoride

Ethylene
Ethylene Dichloride
Fluoromethane
Freon 11, 12, 13, 14, 21, 22, 116, 142b, 143, 151a, 152a
Hexafluoroethane
Liquified Petroleum Gas
Methane
Methyl Chloride
Methylene Chloride
Methylene Fluoride
Nitrogen
Propane
Propylene
Tetrafluoroethylene
Tetrafluoromethane
Trichlorofluoromethane
Trifluoroethane
Trifluoromethane
Vinyl Fluoride
Vinylidene Chloride
Vinylidene Fluoride

Your Goal is: To remove all chemical from contact with the victim's skin as quickly as possible.

SPECIAL WARNING: These chemicals vaporize quickly and present an inhalation hazard as well. Some of them are flammable and explosive. Dispose of contaminated clothing with care.

1. Remove the victim from the source of contamination and far from any fire or smoke.

2. Remove clothing, shoes and jewelry from the affected area, cutting items off if necessary.

3. Wash affected areas with cold water and soap under a shower or running water until all trace of the chemical is gone.

4. Dry the skin gently with a clean, soft towel.

5. Notify a physician, emergency room, or poison control center and inform them of the nature of the substance and the accident.

If the skin is inflamed, painful or hard and white.

6. Call the Emergency Medical Service and arrange for transport to the nearest medical facility.

DO NOT apply any oily substance to the skin.

DO NOT use hot or tepid water to rinse.

DO NOT rub or apply pressure to affected skin.

Acetone
Acrylamide
Aliphatic Alcohols—Amyl
Aliphatic Alcohols—Butyl
Alkanes (liquids/solids)
Allyl Alcohol
Allyl Chloride
Allyl Glycidyl Ether
Allyl Propyl Disulfide
Amyl Acetate
Aniline
Anisidines (ortho)
Anisidines (para)
Asphalt Fumes
Benzene
Benzidine
Bis (Chloromethyl) Ether
Bromoform
Butanol
Butyl Acetate
Butyl Glycidyl Ether (n-)
Butyltoluene
Carbon Tetrachloride
Chloro-1-Nitropropane (1-)
Chloroacetophenone (2-)
Chlorobenzene
Chlorobenzylidene Malonitrile
Chlorobromomethane
Chloroform
Chloronaphthalenes
Chloropentafluoroethane
Chlorophenoxy Compounds
Chloropicrin
Chloropropane
Chloropropene
Cumene
Cyclohexane
Cyclohexanol
Cyclohexanone
DNBP
DNOC
Decane
Decanol
Diacetone Alcohol

Diazomethane
Diborane
Dibutyl Phthalate
Dibutyllead
Dichlorobenzene
Dichloroethylene
Dichloropropane
Dichlorotetrafluoroethane
Diepoxybutane
Diethylaminoethanol
Diethylene Glycol
Diethyllead
Diglycidyl Ether
Diisobutylcarbinol
Dimethylaniline
Dinitrobenzene
Dinitrocresols
Dinitrophenols
Dinitrotoluene
Dioxane
Diphenyl
Diphenylamine
Dipyridyl Chloride
Dipyridyl Dimethyl Sulfate
Diquat
Epichlorohydrin
Ethyl Acetate
Ethyl Acrylate
Ethyl Alcohol
Ethyl Nitrate
Ethylbenzene
Ethylene Chlorohydrin
Ethylene Glycol
Ethylene Glycol Dinitrate
Ethylene Glycol Monomethyl Ether
Ethylene Oxide
Ethyleneimine
Ethylhexyl Acetate
Freon 112, 113, 114, 115
Furfural
Furfuryl Alcohol
Gasoline
Glutaraldehyde
Glycerin

Glycidol
Glycidyl Acrylate
Gramoxone
Halothane
Heptane
Heptanol
Hexachlorobenzene
Hexachloroethane
Hexane
Hexanol
Hydrazine
Hydroquinone
Isobutyl Acetate
Isopropyl Acetate
Isopropyl Alcohol
Ketene
Lead Oleate
Lead Phenate
Lead Phthalate
Lead Stearate
Methyl Acetate
Methyl Acrylate
Methyl Alcohol
Methyl Bromide
Methyl n-Butyl Ketone
Methyl Ethyl Ketone
Methyl Isoamyl Ketone
Methyl Isobutyl Ketone
Methyl Isocyanate
Methyl Isopropyl Ketone
Methyl Methacrylate Monomer
Methyl Nitrate
Methylchloroform
Methylenebis (Phenyl Isocyanate)
Monomethylhydrazine
Naphthalene
Naphthylamines
Naptha
Nickel (fumes and dust)
Nickel Carbonyl
Nitroanilines
Nitrobenzene
Nitrochlorobenzene (p-)
Nitroglycerin

Nitromethane
Nitrocresolic Herbicides
Nitrophenolic Herbicides
Nitrophenols
Nitrotoluene
Nonane
Octane
Paraquat
Pentachloroethane
Pentachlorophenate
Pentachlorophenol
Pentane
Pentanol
Perchloromethyl Mercaptan
Petroleum Ethers
Phenylhydrazine
Phenylhydroxylamine
Phenylnaphthylamine
Picric Acid
Polybrominated Biphenyls (PBBs)
Polychlorinated Biphenyls (PCBs)
Propyl Acetate
Propyl Alcohol
Propyl Nitrate
Propylene Glycol
Propylene Glycol Monomethyl Ether
Propylene Oxide
Pyrethrins
Quaternary Ammonium Compounds
Stoddard Solvent
Styrene
Tetrachlorodifluoroethane
Tetrachloroethane
Tetrachloroethylene
Tetraethyllead
Tetramethyllead
Tetranitromethane
Tetryl
Tolidine (o-)
Toluene
Toluene 2,4-di-Isocyanate
Toluene 2,6-di-Isocyanate
Toluidine
Tributyllead

Trichloroethane
Trichloroethylene
Trichlorotrifluoroethane
Triethylene Glycol
Triethyllead
Trimethyllead
Trinitrobenzene

Trinitrotoluene
Turpentine
Vinyl Acetate
Vinyl Chloride
Xylene
Xylidine

Your Goal is: To remove all chemical from contact with the victim's skin as quickly as possible.

SPECIAL WARNING: These chemicals vaporize easily and present an inhalation hazard as well. Many of them are flammable and explosive. Dispose of contaminated clothing with care.

1. Remove the victim from the source of contamination.

2. Remove clothing, shoes, socks and jewelry from the affected areas. Be careful not to get any of the chemical on your skin or clothing.

3. Wash the affected area under tepid running water using a mild soap.

4. Thoroughly rinse the affected area with tepid water.

5. Dry the skin gently with a clean, soft towel.

6. Notify a physician, emergency room, or poison control center and inform them of the nature of the substance and the accident.

7. Call the Emergency Medical Service and arrange for transport to the nearest medical facility.

8. Keep the victim quiet and arrange to have someone stay with him until medical personnel arrive.

9. Monitor the victim's heartbeat by taking his pulse every few minutes. If the heartbeat is irregular, be prepared to administer CPR.

Aluminum (dust)
Aluminum Hydrate
Aluminum Hydroxide
Aluminum Oxide
Asbestos
Calcium Carbonate
Carbon
Carbon Black
Disodium Phosphate
Fibrous Glass
Kaolin
Magnesium Chloride
Magnesium Sulfate
Polyvinyl Chloride
Silica
Sodium Sulfate
Sodium Thiosulfate
Talc
Titanium (dust and fumes)
Titanium Dioxide
Tungsten Carbide
Yttrium and Compounds

Your Goal is: To remove all chemicals from contact with the victim's skin.

1. Remove the victim from the source of contamination.
2. Remove contaminated clothing. Wet the clothing first to keep down the dust.
3. Wash affected areas with soap and water.
4. Rinse carefully.
5. Dry gently.
6. Get clean, dry clothes for the victim.

DO NOT shake or blow dust off clothing or the body.

8

First Aid in Case of Skin Contact with:

Acetone Cyanohydrin	Ferrocyanides
Acetonitrile	Hydrocyanic Acid
Acrylonitrile	Isobutyronitrile
Adiponitrile	Malononitrile
Bitter Almond Oil (Amygdalin)	Methacrylonitrile
Cherry Laurel Water	Nitroferricyanides (salts)
Cyanogen Bromide	Potassium Cyanide
Cyanogen Chloride	Sodium Cyanide
Cyanogen Iodide	Sodium Thiocyanate
Ferricyanides	Tetramethyl Succinonitrile

Your Goal is: To remove all chemical from contact with the victim's skin as quickly as possible. A delay of only seconds may increase the injury.

1. Remove the victim from the source of contamination and take him **IMMEDIATELY** to the nearest shower or source of clean water.

2. Remove clothing, shoes, socks and jewelry from the affected areas as quickly as possible, cutting them off if necessary. Be careful not to get any of the chemical on your skin or clothing. Wear a respirator approved for cyanide exposure.

3. Wash the affected area under tepid running water using a mild soap.

4. Thoroughly rinse the affected area with tepid water.

5. Dry the skin gently with a clean, soft towel.

6. Notify a physician, emergency room, or poison control center and inform them of the nature of the substance and the accident.

7. Arrange for transport to the nearest medical facility.

8. Do not leave the victim alone. Watch for signs of systemic toxicity. (See INHALATION, Yellow Section 8.)

If the skin is inflamed or painful (particularly with Acrylonitrile):

9. Put the painful part in cold water or apply cold wet dressings on the burned area.

First Aid in Case of Skin Contact with:

Ammonium Carbonate	Potassium Carbonate
Calcium Hydroxide	Sodium Carbonate
Lithium Carbonate	Sodium Silicate
Lithium Hydride	Trisodium Phosphate
Milk of Lime	

Your Goal is: To remove all chemical from contact with the victim's skin as quickly as possible.

1. Remove the victim from the source of contamination and take him **IMMEDIATELY** to the nearest shower or source of clean water.

2. Remove clothing, shoes, socks and jewelry from the affected areas as quickly as possible.

3. Wash the affected area under tepid running water.

4. Rinse the affected area with tepid water.

5. Dry the skin gently with a clean, soft towel.

6. Notify a physician, emergency room, or poison control center and inform them of the nature of the substance and the accident.

7. Dress the victim in clean clothes or cover him with blankets.

If the skin is inflamed or painful:

8. Put the painful part in cold water or apply cold wet dressings on the burned area.

9. Elevate the affected area above the level of the victim's heart.

DO NOT use soap.

DO NOT rub or apply pressure to the affected area.

DO NOT apply any oily substance to the affected skin.

DO NOT use hot water.

10

First Aid in Case of Skin Contact with:

Aluminum Alkyls
Diethylaluminum Chloride
Diethylaluminum Hydride

Triethylaluminum
Triisobutylaluminum
Trimethylaluminum

Your Goal is: To remove all chemical from contact with the victim's skin as quickly as possible. A delay of only seconds may increase the injury.

1. Do not let the victim run away.

2. Very quickly and without touching the victim, wash him down with large amounts of cold water from a hand-held hose, as if to flush away the chemical. **CAUTION:** Do not spray him from the front. The flames will increase in intensity when water is first applied, but will quickly die out.

3. Notify emergency medical personnel of the nature and extent of the injury and arrange for immediate transport to a medical facility.

4. Lay the victim flat on his back on a stretcher without removing the burnt clothing. Turn his head to the side.

5. Cover him with a sterile sheet if available, or a clean dry cloth.

If the victim shows signs of shock:

6. Cover the victim with a blanket.

7. Elevate his feet.

DO NOT touch the victim with your bare hands.

DO NOT handle contaminated articles with your bare hands.

DO NOT break open blisters or remove skin.

DO NOT rub or apply pressure to the affected skin.

DO NOT apply any oily substance to the affected skin.

DO NOT use hot water.

11

First Aid in Case of Skin Contact with:

Aldicarb	OMPA
Carbamates	Organophosphate Compounds
Chlorthion	Paraoxon
DDVP	Parathion
Demeton	Phorate
Diazinon	Phosdrin
Dipterex	Phosphoric Ester
EPN	Ronnel
Isopestox	Sulfotepp
Leptophos	TEPP
Malathion	Trithion
Methyl Parathion	

Your Goal is: To remove all chemical from contact with the victim's skin as quickly as possible. A delay of only seconds may increase the injury.

1. Remove the victim from the source of contamination and take him **IMMEDIATELY** to the nearest shower or source of clean water.

2. Have someone call the Emergency Medical Service and arrange for transport to the nearest medical facility.

3. While wearing polyvinyl or rubber gloves, air tight safety goggles and a respirator approved for organophosphate pesticides, remove clothing, shoes, socks and jewelry from the affected areas as quickly as possible, cutting them off if necessary. Be careful not to get any of the chemical on your skin or clothing. Try not to contaminate healthy skin or eyes with the run off.

4. In the case of extensive splashing of the product, wash the victim down under a cold or luke-warm shower or a hand-held hose, at the same time protecting his eyes. If he is already wearing air tight safety goggles, do not remove them.

5. Wash the affected area under tepid running water using a mild soap. Wash hair and under fingernails and toenails if contaminated.

6. Dry the skin gently with a clean, soft towel.

7. Do not leave the victim alone while awaiting the arrival of medical personnel.

8. Check pulse periodically. Be prepared to give CPR, especially if the pulse is less than 60 beats per minute.

If the victim is experiencing difficulty in breathing:

9. Lay him on his back with his shoulders elevated.

10. Give him oxygen to inhale, if available. (See Inhalation, Yellow Pages Section 3.)

First Aid in Case of Eye Contact with:

Acetaldehyde
Acetic Acid
Acetic Anhydride
Acrolein
Ammonium Sulfide
Antimony and Compounds
Arsenic Trichloride
Arsenicals
Arsine
Barium Fluoride
Benzyl Chloride
Boron Trifluoride
Bromine
Butyraldehyde
Carbon Disulfide
Chlorine
Chlorine Dioxide
Chlorine Trifluoride
Chloroacetaldehyde
Chloroacetic Acid
Creosote
Cresols
Crotonaldehyde
Dichloro-5,5-Dimethylhydantoin
Dimethyl Sulfate
Ethyl Chloroformate
Fluorine
Fluosilicic Acid
Formaldehyde
Formic Acid
Hydriodic Acid
Hydrochloric Acid
Hydrofluoric Acid
Hydrogen Bromide
Hydrogen Chloride
Hydrogen Peroxide
Hydrogen Selenide
Hydrogen Sulfide
Iodine
Isobutyraldehyde
Maleic Anhydride
Methyl Chloroformate

Methyl Mercaptan
Nitric Acid
Nitric Oxide
Nitrogen Dioxide
Nitrogen Trifluoride
Osmic Acid
Oxalic Acid
Oxygen Difluoride
Ozone
Peracetic Acid
Perchloric Acid
Perchloryl Fluoride
Phenol
Phenylenediamine (p-)
Phosgene
Phosphine
Phosphoric Acid
Phosphorus
Phosphorus Chlorides
Phosphorus Pentachloride
Phosphorus Pentasulfide
Phosphorus Trichloride
Potassium Chlorite
Potassium Fluoride
Potassium Fluosilicate
Propionaldehyde
Quinone
Resorcinol
Selenium Hexafluoride
Silane
Sodium Chlorite
Sodium Fluoride
Sodium Fluosilicate
Stibine
Sulfur Dioxide
Sulfur Trioxide
Sulfuric Acid
Sulfurous Acid
Tellurium Hexafluoride
Tributyl Phosphate
Trichloroacetic Acid

Your Goal is: To remove all the chemical from the eye(s) quickly.

1. Remove the victim from the source of contamination and take him to the nearest eye wash, shower, or other source of clean water.

2. Immediately but gently brush, blot or wipe away any liquid or powdered chemical remaining on the face, being careful not to get it on your skin.

3. Gently rinse the affected eye(s) with clean, lukewarm water for at least 15 minutes. Have the victim lie or sit down and tilt his head back. Hold the eyelid(s) open and pour water slowly over the eyeball(s) at the inner corners, letting the water run out of the outer corners.

   The victim may be in great pain and want to keep his eyes closed but you must rinse the chemical out of his eye(s) in order to prevent permanent damage.

4. Ask the victim to look up, down and side to side as you rinse in order to better reach all parts of the eye(s). Have the victim remove contact lenses if he is wearing them and continue rinsing.

5. Arrange for transport to the nearest medical facility for examination and treatment by a physician as soon as possible. Tell the Emergency Medical Service personnel the name of the chemical and the nature of the accident.

   Even if there is no pain and vision is good, a physician should still examine the eye(s) since delayed damage may occur.

NOTE: If the eye(s) is splashed with **fluorine compound** or sprayed with fluorine gas, several drops of sterile calcium gluconate 10% solution in the eye(s) after rinsing is helpful if this is available. Consult medical personnel before doing this.

DO NOT let the victim rub his eye(s).

DO NOT let the victim keep his eyes tightly shut.

DO NOT introduce oil or ointment into the eye(s) without medical advice.

DO NOT use hot water.

Aliphatic Amines
Ammonia
Ammonium Hydroxide
Butylamine
Calcium Carbide
Calcium Oxide
Cement
Dibutylamine
Diethylamine
Dimethylamine
Dipropylamine
Ethanolamine
Ethylamine
Isopropylamine

Lime
Methylamine
Milk of Lime
Potassium
Potassium Hydroxide
Potassium Oxide
Propylamine
Sodium
Sodium Hydroxide
Sodium Oxide
Sodium Peroxide
Triethylamine
Trimethylamine

Your Goal is: To remove all the chemical from the eye(s) quickly.

SPECIAL WARNING: Some of these solids may get very hot when they contact water.

1. Remove the victim from the source of contamination and take him to the nearest eye wash, shower, or other source of clean water.

2. Immediately but gently brush, blot or wipe away any liquid or powdered chemical remaining on the face, being careful not to get it on your skin.

3. Gently rinse the affected eye(s) with clean, lukewarm water for at least 15 minutes. Have the victim lie or sit down and tilt his head back. Hold the eyelid(s) open and pour water slowly over the eyeball(s) at the inner corners, letting the water run out of the outer corners.

   The victim may be in great pain and want to keep his eyes closed but you must rinse the chemical out of his eye(s) in order to prevent permanent damage.

4. Ask the victim to look up, down and side to side as you rinse in order to better reach all parts of the eye(s). Have the victim remove contact lenses if he is wearing them.

5. During rinsing, check to make sure that no solid particles of the chemical remain in the creases of the eye(s) or on the lashes and brows. Rinse away if they do.

6. Arrange for transport to the nearest medical facility for examination and treatment by a physician as soon as possible. Tell the Emergency Medical Service personnel the name of the chemical and the nature of the accident.

7. Continue rinsing the eye(s) until medical help arrives, even if more than 15 minutes has elapsed.

NOTE: Even if there is no pain and vision is good, a physician should still examine the eye(s) since delayed damage may occur.

NO NOT let the victim rub his eye(s).

DO NOT let the victim keep his eyes tightly shut.

DO NOT introduce oil or ointment into the eye(s) without medical advice.

DO NOT use hot water.

3

First Aid in Case of Eye Contact with:

Ammonium Carbonate
Aniline
Anisidines (ortho)
Anisidines (para)
Benzidine
Calcium Hydroxide
Chlorophenoxy Compounds
DNBP
DNOC
Dimethylaniline
Dinitrobenzene
Dinitrocresols
Dinitrophenols
Dinitrotoluene
Ethyl Nitrate
Ethylene Glycol Dinitrate
Hydroquinone
Lithium Carbonate
Lithium Hydride
Methyl Nitrate
Monomethylhydrazine
Naphthalene
Naphthylamines
Nitroanilines

Nitrobenzene
Nitrochlorobenzene (p-)
Nitroglycerin
Nitrocresolic Herbicides
Nitrophenolic Herbicides
Nitrophenols
Nitrotoluene
Pentachlorophenate
Pentachlorophenol
Phenylhydrazine
Phenylhydroxylamine
Phenylnaphthylamine
Potassium Carbonate
Propyl Nitrate
Sodium Carbonate
Sodium Silicate
Tetranitromethane
Tolidine (o-)
Toluidine
Trinitrobenzene
Trinitrotoluene
Trisodium Phosphate
Xylidine

Your Goal is: To remove all the chemical from the eye(s) quickly.

SPECIAL WARNING: Other symptoms may appear because of absorption of the chemicals into the blood stream through the eye(s).

1. Remove the victim from the source of contamination and take him to the nearest eye wash, shower, or other source of clean water.

2. Immediately but gently brush, blot or wipe away any liquid or powdered chemical remaining on the face, being careful not to get it on your skin.

3. Gently rinse the affected eye(s) with clean, lukewarm water for at least 15 minutes. Have the victim lie or sit down and tilt his head back. Hold the eyelid(s) open and pour water slowly over the eyeball(s) at the inner corners, letting the water run out of the outer corners.

The victim may be in great pain and want to keep his eyes closed but you must rinse the chemical out of his eye(s) in order to prevent permanent damage.

4. Ask the victim to look up, down and side to side as you rinse in order to better reach all parts of the eye(s). Have the victim remove contact lenses if he is wearing them.

5. During the rinsing, check to make sure that no solid particles of the chemical remains in the creases of the eye(s) or on the lashes and brows. Rinse away if they do.

6. If pain persists, continue washing the eye(s) for another 15 minutes or until pain ceases.

7. Arrange for transport to the nearest medical facility for examination and treatment by a physician as soon as possible. Tell the Emergency Medical Service personnel the name of the chemical and the nature of the accident.

Even if there is no pain and vision is good, a physician should still examine the eye(s) since delayed damage may occur.

DO NOT let the victim rub his eye(s).

DO NOT let the victim keep his eyes tightly shut.

DO NOT introduce oil or ointment into the eye(s) without medical advice.

DO NOT use hot water.

4

First Aid in Case of Eye Contact with:

Acetylene
Alkanes (gasses, C_1 to C_4)
Butadiene
Butane
Carbon Dioxide
Carbon Dioxide Snow
Chlorodifluoroethane
Chlorodifluoromethane
Chloroethane
Chlorofluoroethane
Chlorofluoromethane
Chloromethane
Chlorotrifluoroethylene
Chlorotrifluoromethane
Dichlorodifluoromethane
Dichloroethane
Dichlorofluoromethane
Difluoroethanes
Difluoroethylene
Ethane
Ethyl Ether
Ethyl Fluoride

Ethylene
Ethylene Dichloride
Fluoromethane
Freon 11, 12, 13, 14, 21, 22, 116, 142b, 143, 151a, 152a
Hexafluoroethane
Liquefied Petroleum Gas
Methane
Methyl Chloride
Methylene Chloride
Methylene Fluoride
Nitrogen
Propane
Propylene
Tetrafluoroethylene
Tetrafluoromethane
Trichlorofluoromethane
Trifluoroethane
Trifluoromethane
Vinyl Fluoride
Vinylidene Chloride
Vinylidene Fluoride

Your Goal is: To remove all the chemical from the eye(s) quickly.

SPECIAL WARNING: Freezing may occur upon contact with these highly volatile chemicals.

1. Remove the victim from the source of contamination and take him to the nearest eye wash, shower, or other source of clean water.

2. Open the eyelid(s) wide to let the chemical evaporate.

3. Gently rinse the affected eye(s) with clean, cool water for at least 15 minutes. Have the victim lie or sit down and tilt his head back. Hold the eyelid(s) open and pour water slowly over the eyeball(s) at the inner corners, letting the water run out of the outer corners.

   The victim may be in great pain and want to keep his eyes closed but you must rinse the chemical out of his eye(s) in order to prevent permanent damage.

4. Ask the victim to look up, down and side to side as you rinse in order to better reach all parts of the eye(s). Have the victim remove contact lenses if he is wearing them and continue rinsing.

5. Arrange for transport to the nearest medical facility for examination and treatment by a physician as soon as possible. Tell the Emergency Medical Service personnel the name of the chemical and the nature of the accident.

   Even if there is no pain and vision is good, a physician should still examine the eye(s) since delayed damage may occur.

6. If the victim cannot tolerate light, protect his eye(s) with a clean, loosely tied handkerchief or strip of cloth or bandage. Be sure to maintain verbal communication and physical contact with the victim.

DO NOT let the victim rub his eye(s).

DO NOT let the victim keep his eyes tightly shut.

DO NOT introduce oil or ointment into the eye(s) without medical advice.

DO NOT use hot or even tepid water.

First Aid in Case of Eye Contact with:

Acetone
Acetone Cyanohydrin
Acetonitrile
Acrylamide
Acrylonitrile
Adiponitrile
Aldrin
Aliphatic Alcohols—Amyl
Aliphatic Alcohols—Butyl
Allyl Alcohol
Allyl Chloride
Allyl Glycidyl Ether
Allyl Propyl Disulfide
Aminopyridine
Amyl Acetate
Benzene
Bis(Chloromethyl) Ether
Bitter Almond Oil (Amygdalin)
Bromoform
Butanol
Butyl Acetate
Butyl Glycidyl Ether (n-)
Butyltoluene
Camphor
Carbon Tetrachloride
Cherry Laurel Water
Chlordane
Chloro-1-Nitropropane (1-)
Chloroacetophenone (2-)
Chlorobenzene
Chlorobenzylidene Malonitrile
Chlorobromomethane
Chloroform
Chloropentafluoroethane
Chloropicrin
Chloropropane
Chloropropene
Cumene
Cyanogen Bromide
Cyanogen Chloride
Cyanogen Iodide
Cyclohexane
Cyclohexanol
Cyclohexanone

Decaborane
Decane
Decanol
Diacetone Alcohol
Diazomethane
Diborane
Dibutyl Phthalate
Dichlorobenzene
Dichloroethylene
Dichloropropane
Dichlorotetrafluoroethane
Diepoxybutane
Diethylaminoethanol
Diethylene Glycol
Diglycidyl Ether
Diisobutylcarbinol
Dimethylhydrazine (1,1-)
Dioxane
Dipyridyl Chloride
Dipyridyl Dimethyl Sulfate
Diquat
Epichlorohydrin
Ethyl Acetate
Ethyl Acrylate
Ethyl Alcohol
Ethylbenzene
Ethylene Chlorohydrin
Ethylene Glycol
Ethylene Glycol Monomethyl Ether
Ethylene Oxide
Ethyleneimine
Ethylhexyl Acetate
Ferricyanides
Ferrocyanides
Freon 112, 113, 114, 115
Furfural
Furfuryl Alcohol
Gasoline
Glutaraldehyde
Glycerin
Glycidol
Glycidyl Acrylate
Gramoxone
Halothane

Heptane
Heptanol
Hexachloroethane
Hexane
Hexanol
Hydrazine
Hydrocyanic Acid
Isobutyl Acetate
Isobutyronitrile
Isoproyl Acetate
Isopropyl Alcohol
Ketene
Lindane
Malononitrile
Methacrylonitrile
Methyl Acetate
Methyl Acrylate
Methyl Alcohol
Methyl Bromide
Methyl n-Butyl Ketone
Methyl Ethyl Ketone
Methyl Isoamyl Ketone
Methyl Isobutyl Ketone
Methyl Isocyanate
Methyl Isopropyl Ketone
Methyl Methacrylate Monomer
Methylchloroform
Methylenebis (Phenyl Isocyanate)
Naptha
Nickel (fumes and dust)
Nickel Carbonyl
Nitroferricyanides (salts)
Nitromethane
Nonane
Octane
Organochlorines

Paraquat
Pentaborane
Pentachloroethane
Pentane
Pentanol
Perchloromethyl Mercaptan
Petroleum Ethers
Picric Acid
Potassium Cyanide
Propyl Acetate
Propyl Alcohol
Propylene Glycol
Propylene Glycol Monomethyl Ether
Propylene Oxide
Pyrethrins
Quaternary Ammonium Compounds
Sodium Cyanide
Sodium Thiocyanate
Stoddard Solvent
Styrene
Tetrachlorodifluoroethane
Tetrachloroethane
Tetrachloroethylene
Tetramethyl Succinonitrile
Tetryl
Toluene
Toluene 2,4-di-Isocyanate
Toluene 2,6-di-Isocyanate
Trichloroethane
Trichloroethylene
Trichlorotrifluoroethane
Triethylene Glycol
Turpentine
Vinyl Acetate
Vinyl Chloride
Xylene

Your Goal is: To remove all the chemical from the eye(s) quickly.

1. Remove the victim from the source of contamination and take him to the nearest eye wash, shower, or other source of clean water.

2. Gently rinse the affected eye(s) with clean, lukewarm water for at least

15 minutes. Have the victim lie or sit down and tilt his head back. Hold the eyelid(s) open and pour water slowly over the eyeball(s) at the inner corners, letting the water run out of the outer corners.

3. Ask the victim to look up, down and side to side as you rinse in order to better reach all parts of the eye(s). Have the victim remove contact lenses if he is wearing them.

4. Arrange for transport to the nearest medical facility for examination and treatment by a physician as soon as possible. Tell the Emergency Medical Service personnel the name of the chemical and the nature of the accident.

   Even if there is no pain and vision is good, a physician should still examine the eye(s) since delayed damage may occur.

5. If the victim cannot tolerate light, protect his eye(s) with a clean, loosely tied handkerchief or strip of clean, soft cloth or bandage. Be sure to maintain verbal communication and physical contact with the victim.

DO NOT let the victim rub his eye(s).

DO NOT let the victim keep his eyes tightly shut.

DO NOT introduce oil or ointment into the eye(s) without medical advice.

DO NOT use hot water.

First Aid in Case of Eye Contact with:

Alkali Dichromates
Alkali Meta-Borates
Alkanes (liquids/solids)
Aluminum Chloride
Aluminum Trichloride
Ammonium Chlorate
Ammonium Perchlorate
Arsenic
Asphalt Fumes
Barium (soluble salts)
Barium Acetate
Barium Carbonate
Barium Chloride
Barium Hydroxide
Barium Nitrate
Barium Oxide
Barium Sulfide
Boric Acid
Cadmium (dust and fumes) (metal)
Calcium Chloride
Calcium Dichromate
Calcium Hypochlorite
Caprolactam
Chlorinated Lime
Chloronaphthalenes
Chromic Acid
Chromium Chloride
Copper Chloride
Copper Sulfate
Dibutyllead
Dibutyltin
Diethyllead
Diethylmercury
Diethyltin
Dihexyltin
Diiododiethyltin
Dimethylmercury
Dimethyltin
Dioctyltin
Diphenyl
Diphenylamine
Disodium Phosphate
Dithiocarbamates
Ethylmercuric Chloride

Ethylmercuric Hydroxide
Ethylmercury
Hexachlorobenzene
Iron Chloride
Lead (dust and fumes)
Lead Acetate
Lead Antimonate
Lead Arsenate
Lead Carbonate
Lead Chromate
Lead Chromate (yellow)
Lead Dioxide
Lead Nitrate
Lead Oleate
Lead Oxide (PbO)
Lead Oxide (red)
Lead Oxychloride
Lead Phenate
Lead Phthalate
Lead Stearate
Lead Subacetate
Lead Sulfide
Magnesium Chloride
Magnesium Sulfate
Mercuric Chloride
Mercuric Iodide (red)
Mercurous Chloride
Mercurous Iodide
Mercury (metal)
Mercury (organic compounds)
Mercury (soluble salts)
Mercury Acetate
Mercury Fulminate
Mercury Nitrate
Mercury Oxycyanide
Methylmercury
Methylmercury Borate
Methylmercury Hydroxide
Methylmercury Iodide
Methylmercury Nitrate
Methylmercury Phosphate
Paraffins
Perborates
Phenylmercuric Acetate

Phenylmercury
Phenylmercury Oleate
Phthalic Anhydride
Platinum and Compounds
Polybrominated Biphenyls (PBBs)
Polychlorinated Biphenyls (PCBs)
Potassium Chlorate
Potassium Chloride
Potassium Chromate
Potassium Dichromate
Potassium Perchlorate
Sodium Bicarbonate
Sodium Borate
Sodium Chlorate
Sodium Chloride
Sodium Chromate
Sodium Dichromate
Sodium Hypochlorite
Sodium Perchlorate
Sodium Sulfate
Sodium Thiosulfate

Tetrabutyltin
Tetraethyllead
Tetraethyltin
Tetraisoalkyltin
Tetramethyllead
Tetrapentyltin
Tetrapropyltin
Thiocarbamates
Titanium Chlorides
Tributyllead
Tributyltin
Triethyllead
Trimellitic Anhydride
Trimethyllead
Trimethyltin
Triphenyltin
Tripropyltin
Uranium and Compounds
Vanadium and Compounds
Zinc Chloride

Your Goal is: To remove all the chemical from the eye(s) quickly.

1. Remove the victim from the source of contamination and take him to the nearest eye wash, shower, or other source of clean water.

2. Gently brush, blot or wipe away any liquid or powdered chemical remaining on the face.

3. Gently rinse the affected eye(s) with clean, lukewarm water for at least 15 minutes. Have the victim lie or sit down and tilt his head back. Hold the eyelid(s) open and pour water slowly over the eyeball(s) at the inner corners, letting the water run out of the outer corners.

4. Ask the victim to look up, down and side to side as you rinse in order to better reach all parts of the eye(s).

5. During the rinsing, check to make sure that no solid particles of the chemical remain in the creases of the eye(s) or on the lashes and brows. Rinse away if they do. Have the victim remove contact lenses if he is wearing them.

6. Arrange for transport to the nearest medical facility for examination and treatment by a physician as soon as possible. Tell the Emergency Medical Service personnel the name of the chemical and the nature of the accident.

   Even if there is no pain and vision is good, a physician should still examine the eye(s) since delayed damage may occur.

DO NOT let the victim rub his eye(s).

DO NOT let the victim keep his eyes tightly shut.

DO NOT introduce oil or ointment into the eye(s) without medical advice.

DO NOT use hot water.

7

First Aid in Case of Eye Contact with:

Aluminum (dust)
Aluminum Hydrate
Aluminum Hydroxide
Aluminum Oxide
Asbestos
Calcium Carbonate
Carbon
Carbon Black
Fibrous Glass

Kaolin
Polyvinyl Chloride
Silica
Talc
Titanium (dust and fumes)
Titanium Dioxide
Tungsten Carbide
Yttrium and Compounds

Your Goal is: To remove the dust from the victim's eye(s).

1. Remove the victim from the source of contamination and take him to the nearest eye wash, shower, or other source of clean water.

2. Gently brush, or wipe away any powdered chemical remaining on the face. Have the victim remove contact lenses if he is wearing them.

3. Gently rinse the affected eye(s) with clean, lukewarm water until the pain is gone for 2 or 3 minutes. Have the victim lie or sit with head back. Hold the eyelid(s) open and pour water slowly over the eyeball(s) at the inner corners, letting the water run out the outer corners.

4. Ask the victim to look up, down and side to side as you rinse in order to better reach all parts of the eye(s).

5. Arrange for transport to the nearest medical facility for examination and treatment by a physician.

DO NOT let the victim rub his eye(s).

DO NOT introduce oil or ointment into the eye(s) without medical advice.

8

First Aid in Case of Eye Contact with:

Aluminum Alkyls	Triethylaluminum
Diethylaluminum Chloride	Triisobutylaluminum
Diethylaluminum Hydride	Trimethylaluminum

Your Goal is: To extinguish the flames and remove all the chemical from the eye(s) and face.

SPECIAL WARNING: These materials ignite and burn on contact with water.

1. Make sure the victim does not run away.

2. Flush the victim's head immediately with large amounts of water using a hose if possible. (The flames will increase in intensity when the water is first applied but will quickly die out).

3. Lay the victim down on a **DRY** flat surface or stretcher. Cover with a blanket to prevent chilling. You may elevate the head if he is more comfortable in this position. If he feels like vomiting, turn his head to the side.

4. Put a clean soft cloth or bandage over his eye(s). Be sure to maintain verbal communication and physical contact with the victim.

5. Arrange for transport to the nearest medical facility for examination and treatment by a physician as soon as possible.

   Tell the Emergency Medical Service personnel the name of the chemical and the nature of the accident.

DO NOT let the victim rub his eye(s).

DO NOT introduce oil or ointment into the eye(s) without medical advice.

9

First Aid in Case of Eye Contact with:

Aldicarb
Carbamates
Chlorthion
DDVP
Demeton
Diazinon
Dipterex
EPN
Isopestox
Leptophos
Malathion
Methyl Parathion

OMPA
Organophosphate Compounds
Paraoxon
Parathion
Phorate
Phosdrin
Phosphoric Ester
Ronnel
Sulfotepp
TEPP
Trithion

Your Goal is: To remove all the chemical from the eye(s) quickly.

SPECIAL WARNING: Dangerous amounts of these chemicals can be absorbed into the bloodstream as a result of eye and skin contact with small (1 or 2 teaspoon) amounts. Watch for other symptoms and refer to the section on skin contact.

1. Remove the victim from the source of contamination and take him to the nearest eye wash, shower, or other source of clean water.

2. Quickly brush, blot or wipe away any liquid or powdered chemical remaining on the face, being careful not to get it on your skin.

3. Gently rinse the affected eye(s) with clean, lukewarm water for at least 15 minutes. Have the victim lie or sit down and tilt his head back. Hold the eyelid(s) open and pour water slowly over the eyeball(s) at the inner corners, letting the water run out of the outer corners.

4. Ask the victim to look up, down and side to side as you rinse in order to better reach all parts of the eye(s). Have him remove contact lenses if he is wearing them.

5. Arrange for transport to the nearest medical facility for examination and treatment by a physician as soon as possible. Tell the Emergency Medical Service personnel the name of the chemical and the nature of the accident.

   Even if there is no pain or other symptoms and vision is good, a physician should still examine the victim since delayed symptoms may occur.

If the victim is conscious but has difficulty in breathing:

6. Lay him on his back with his head lowered and his legs raised.

7. Loosen his collar and belt.

8. Cover him with a blanket.

9. If available, give oxygen until Emergency Medical Service personnel arrive.

In the case of the above symptoms, the Emergency Medical Service should be called immediately to treat the victim. The drug atropine, when given intravenously, can be life-saving.

DO NOT let the victim rub his eye(s).

DO NOT let the victim keep his eyes tightly shut.

DO NOT introduce oil or ointment into the eye(s) without medical advice.

DO NOT use hot water.

Appendix A
General Instructions in Case of
Poisoning
by UNKNOWN Chemicals

It is relatively rare in most situations where chemicals are handled to not know their nature and exact identity. But if no one is available to tell you what happened and you are unfamiliar with the situation, you must act anyway. Remember the Golden Rules:

1. **Quickly protect yourself from exposure before attempting to rescue the victim.**

2. **Decontaminate the victim and terminate exposure.**

3. **Treat cessation of breathing first.**

4. **If the heart is not beating, perform CPR.**

5. **Treat eye injuries next.**

6. **Treat skin contact.**

7. **Treat shock.**

8. **Call for help.**

Make a quick assessment of the likely route of exposure by examining the eyes, mouth, nose, and skin of the victim for signs of the chemical itself or damage it has caused such as swelling, redness, bleeding, burns, discharge of fluid or mucous, or pallor. Drooling, difficulty swallowing, a distended and painful or hard, rigid abdomen all indicate possible ingestion of a corrosive or caustic substance. If respirations are rapid, shallow, noisy or labored, suspect inhalation. If the face has been splashed with chemical, eye contact is likely.

A. Poisoning by inhalation

1. Remove the victim from exposure while protecting yourself from exposure.

2. If breathing has stopped, administer artificial respiration using a bag-valve mask. **DO NOT** use mouth to mouth respiration if the nature of the chemical exposure is unknown. If a bag-valve mask is not available, use the chest pressure-arm lift technique.

3. Maintain an open airway.

4. Notify the Emergency Medical Service of the nature of the accident and arrange for transport to a medical facility.

B. Poisoning by ingestion

1. Remove the victim from exposure while protecting yourself from exposure.

2. Call a poison control center, emergency room or a physician for advice.

3. Notify the Emergency Medical Service of the nature of the accident and arrange for transport to a medical facility.

a. If the victim is conscious:

1. Have him rinse out his mouth with water.

2. Give him 1 or 2 cups of water or milk to drink. If he becomes nauseated or cannot swallow, **STOP.**

3. **DO NOT** induce vomiting if he has abdominal pain, a distended abdomen or a rigid, hard abdomen.

4. **DO NOT** induce vomiting if he has burns in or around his mouth or if he cannot swallow.

Alternatively, if there are no signs of burns, swallowing difficulty or abdominal problems and if so advised by a physician or a poison control center:

5. Induce vomiting by giving 2 teaspoonfuls of syrup of ipecac. Follow with at least 1 cup of water. **DO NOT** use milk. If you do not have syrup of ipecac, induce vomiting by asking the victim to touch the back of his throat with his finger, a spoon handle or a blunt instrument.

6. Have the victim sit up or lean forward while vomiting.

7. Save the vomitus and give it to the Emergency Medical Service personnel to take to the medical facility for analysis.

8. Give the victim 1 or 2 cups of water to drink after vomiting has ceased.

9. Keep talking with him to prevent sleepiness.

b. If the victim is unconscious:

1. Lay him on his left side, bending his right hip.

2. Maintain an open airway.

3. Arrange for transport to the nearest medical facility.

4. Stand by to administer artificial respiration and CPR if needed. Be sure to wipe or rinse all traces of the chemical from in and around the victim's mouth before giving artificial respiration. **DO NOT** use mouth to mouth respiration if the nature of the chemical exposure is unknown. Alternatively, use a bag-valve mask or the chest pressure-arm lift technique.

5. If he vomits, save the vomitus and send it to the medical facility for analysis.

6. If the victim shows signs of shock (a weak, rapid pulse; pale, clammy skin; cold hands and feet) elevate his feet 8 to 12 inches and cover him with a blanket.

DO NOT give an unconscious person anything to drink.

DO NOT give anything to drink to someone who is convulsing.

C. Poisoning by skin contact

1. Remove the victim from the contaminated area, being careful to protect your lungs, skin and eyes.

2. Remove the victim's clothing, shoes, and jewelry from the affected areas, cutting them off if necessary. Do this under a shower or while flushing with water.

3. Continue to flush with water until all trace of the chemical is gone and any feeling of soapiness has disappeared also. Rinse for at least 15 minutes.

4. Cover the victim with a blanket or dry clothing.

5. Notify a physician, emergency room or poison control center of the accident and obtain advice.

   a. In case of inflammation, burns, blisters or pain:

   1. Loosely apply a dry sterile dressing if available or use a clean dry cloth.

   2. Notify the Emergency Medical Service of the nature of the accident and arrange for transport to a medical facility.

   b. If the victim is in a state of shock:

   1. Lay him down on his side and cover him with a blanket.

   2. Elevate his feet.

   3. Notify the Emergency Medical Service of the nature of the accident and arrange for transport to a medical facility.

DO NOT break open blisters or remove skin. If clothing is stuck to the skin after flushing with water, do not remove it.

DO NOT rub or apply pressure to the affected skin.

DO NOT apply any oily substance to the affected skin.

DO NOT use hot water.

D. Poisoning by eye contact

1. Remove the victim from the contaminated area, being careful to protect your lungs, skin and eyes.

2. Act quickly. Seconds count. Flush the victim's eye(s) with clean tepid water for at least 15 minutes. Have the victim lie or sit down and tilt his head back. Hold his eyelid(s) open and pour water slowly over the eyeball(s) starting at the inner corners by the nose and letting the water run out of the outer corners.

   The victim may be in great pain and want to keep his eyes closed or rub them but you must rinse the chemical out of the eye(s) in order to prevent possible permanent damage.

3. Ask the victim to look up, down and side to side as you rinse.

4. Call the Emergency Medical Service and arrange for transport to the nearest medical facility for examination and treatment as soon as possible. Even if there is no pain and vision is good, a physician should examine the eye(s) since delayed damage may occur.

a. If the eye(s) is painful:

1. Cover loosely with gauze or a clean, dry cloth.

2. Maintain verbal and physical contact with the victim.

Appendix B
Meaning of DOT Symbols

The United States Department of Transportation (DOT) requires labeling of all hazardous materials when they are transported. The labels must appear as diamond shaped placards on the front, rear and sides of the vehicle. A four digit number appears on the label which identifies the chemical by name when looked up in the Department of Transportation's *Emergency Response Guide Book for Hazardous Materials Incidents* which is available from the U.S. Government Printing Office. The Chemical Manufacturers Association maintains a toll free "hot line" 24 hours a day, 7 days a week which can also provide help with product identification and emergency assistance. This service is called CHEMTREC. Their toll free number is (800) 424-9300.

DEPARTMENT OF TRANSPORTATION (DOT) LABELS AND THEIR MEANING

If The Placard Says:	Then The Hazard Class Is:	Definition of the Class
Explosives A	Class A Explosive	Detonating or otherwise of maximum hazard.
Explosives B	Class B Explosive	In general, function by rapid combustion rather than detonation and include some explosive devices such as special fireworks, flash powders, etc. *Flammable hazard.*
Dangerous	Class C Explosive	Certain types of manufactured articles containing Class A or Class B explosives, or both, as components but in restricted quantities, and certain types of fireworks. *Minimum hazard.*
Blasting Agents	Blasting Agents	A material designed for blasting that has been tested and found to be so insensitive that there is very little probability of accidental initiation to explosion or of transition from deflagration to detonation.
Combustible Liquid	Combustible Liquid	Any liquid having a flash point above 100°F. and below 200°F as measured by tests in the DOT regulations

If The Placard Says:	Then The Hazard Class Is:	Definition of the Class
Corrosive	Corrosive Material	Any liquid or solid that causes visible destruction of human skin tissue or a liquid that has a severe corrosion rate on steel.
Flammable	Flammable Liquid	Any liquid having a flash point below 100°F as measured by tests in the DOT regulations.
Flammable Gas	Flammable Gas	Any compressed gas meeting the requirements for lower flammability limit, flammability limit range, flame projection, or flame propagation criteria as specified in DOT regulations.
"Name of Gas"	NonFlammable Gas	Any compressed gas other than a flammable compressed gas.
Flammable Solid	Flammable Solid	Any solid material, other than an explosive, that is liable to cause fires through friction, retained heat from manufacturing or processing, or that can be ignited readily and when ignited burns so vigorously and persistently, that it can create a serious transportation hazard.
Organic Peroxide	Organic Peroxide	An organic compound containing the bivalent -O-O structure that may

(*continued*)

If The Placard Says:	Then The Hazard Class Is:	Definition of the Class
Organic Peroxide (*cont.*)	Organic Peroxide (*cont.*)	be considered a derivative of hydrogen peroxide where one or more of the hydrogen atoms have been replaced by organic radicals, must be classed as an organic peroxide unless: it classifies as an explosive A or Explosive B; the predominent hazard of the material is other than from an organic peroxide; the material is forbidden for transport; or the material will not pose a hazard in transport.
Oxidizer	Oxidizer	A substance such as chlorate, permanganate, inorganic peroxide, or a nitrate, that yields oxygen readily to stimulate the combustion of organic matter.
Poison Gas	Poison A	*Extremely Dangerous Poisons*—Poisonous gases or liquids of such nature that a very small amount of the gas, or vapor of the liquid, mixed with air is *dangerous to life*.
Poison	Poison B	*Less Dangerous Poisons*—Substances, liquids, or solids (including pastes and semi-solids), other than Class A or Irritating Materials, that are known to be so toxic to man as

If The Placard Says:	Then The Hazard Class Is:	Definition of the Class
Poison (*cont.*)	Poison B (*cont.*)	to afford a hazard to health during transportation; or that, in the absence of adequate data on human toxicity, are presumed to be *toxic to man.*
Dangerous	Irritating	A liquid or solid substance that upon contact with fire or air gives off dangerous or intensely irritating fumes, but *not including any poisonous material, Class A.*
Etiologic Agent Biomedical Material	Etiologic Agent	A viable micro-organism, or its toxin that causes or may cause human disease.
ORM	ORM-Other Regulated Materials	1. Any material that may pose an unreasonable risk to health and safety or property when transported in commerce; and 2. Does not meet any of the definitions of the other hazard classes specified; or 3. Has been reclassed an ORM (specifically or permissively). **NOTE:** A material with a flashpoint of 100°F to 200°F may not be classed as an ORM if it is a hazardous waste or is offered in a package having a rated capacity of more than 110 gallons.

(*continued*)

If The Placard Says:	Then The Hazard Class Is:	Definition of the Class
ORM-A	ORM-A	A material which has an anesthetic, irritating, noxious, toxic, or other similar property and can cause extreme annoyance or discomfort to passengers and crew in the event of leakage during transportation.
ORM-B	ORM-B	A material (including a solid when wet with water) capable of causing significant damage to a transport vehicle or vessel from leakage during transportation. Materials meeting the following criterion are ORM-B materials: A liquid substance that has a corrosion rate exceeding 0.250 inch per year (IPY) on aluminum (nonclad 7075-T6) at a test temperature of 103°F.
ORM-C	ORM-C	A material that has other inherent characteristics not described as an ORM-A or ORM-B but which make it unsuitable for shipment, unless properly identified and prepared for transportation.
ORM-D	ORM-D	A material such as a consumer commodity that, though otherwise subject to DOT regulations, presents a

If The Placard Says:	Then The Hazard Class Is:	Definition of the Class
ORM-D *(cont.)*	ORM-D *(cont.)*	limited hazard during transportation due to its form, quantity and packaging.
ORM-E	ORM-E	A material that is not included in any other hazard class, but is subject to DOT requirements. Materials in this class include hazardous wastes and hazardous substances.
Spontaneously Combustible	Spontaneously Combustible Material (Solid)	A solid substance (including sludges and pastes) that may undergo spontaneous heating or self-ignition under conditions normally incident to transportation or that may, upon contact with the atmosphere, undergo an increase in temperature and ignite.
Dangerous When Wet	Water Reactive Material (Solid)	Any solid substance (including sludges and pastes) that, by interaction with water, is likely to become spontaneously flammable or give off flammable or toxic gases in dangerous quantities.

Appendix C

Glossary of Commercial and Common Pesticide Names

Name	See
AAtack	Dithiocarbamates and Thiocarbamates
Abate	Organophosphate Compounds
Abathion	Organophosphate Compounds
Acaraben	Organochlorines
Acephate	Organophosphate Compounds
Acrex	Nitrophenolic and Nitrocresolic Herbicides
Actellic	Organophosphate Compounds
Afugan	Organophosphate Compounds
Agritox	Organophosphate Compounds
Agrotec	Chlorophenoxy Compounds
Agrothion	Organophosphate Compounds
Aldicarb	Carbamates
Aldrin	Organochlorines
Aldrite	Organochlorines
Allethrin	Pyrethrins
Amaze	Organophosphate Compounds
Ambox	Nitrophenolic and Nitrocresolic Herbicides
Ambush	Pyrethrins
Aminocarb	Carbamates
Amoxone	Chlorophenoxy Compounds
Ancrack	Nitrophenolic and Nitrocresolic Herbicides
Ansar 170 HC	Arsenicals
Ansar 529 HC	Arsenicals
Ansar 8100	Arsenicals

(continued)

Name	See
Ansar DSMA Liquid	Arsenicals
Anthio	Organophosphate Compounds
Aphox	Carbamates
Aqua-Kleen	Chlorophenoxy Compounds
Aquacide	Paraquat and Diquat
Arasan	Dithiocarbamates and Thiocarbamates
Arrhenal	Arsenicals
Arsenic	Arsenicals
Arsenic Acid	Arsenicals
Arsenic Trioxide	Arsenicals
Arsenious Oxide	Arsenicals
Arsine	Arsenicals
Arsinyl	Arsenicals
Arsonate Liquid	Arsenicals
Aspon	Organophosphate Compounds
Aspor	Dithiocarbamates and Thiocarbamates
Aules	Dithiocarbamates and Thiocarbamates
Avadex	Dithiocarbamates and Thiocarbamates
Avadex-BW	Dithiocarbamates and Thiocarbamates
Azinphos-Methyl	Organophosphate Compounds
Azodrin	Organophosphate Compounds
BH2, 4-D	Chlorophenoxy Compounds
BHC	Organochlorines
Banvel	Chlorophenoxy Compounds
Barthrin	Pyrethrins
Basanite	Nitrophenolic and Nitrocresolic Herbicides
Baygon	Carbamates
Bayrusil	Organophosphate Compounds
Baytex	Organophosphate Compounds
Baythion	Organophosphate Compounds
Belmark	Pyrethrins
Benzene Hexachloride	Organochlorines
Bidrin	Organophosphate Compounds
Binapacryl	Nitrophenolic and Nitrocresolic Herbicides
Bioresmethrin	Pyrethrins
Birlane	Organophosphate Compounds
Black Leaf Grass, Weed, Vegetation Killer Spray	Pentachlorophenol
Bladafum	Organophosphate Compounds
Bo-Ana	Organophosphate Compounds
Bolstar	Organophosphate Compounds
Bomyl	Organophosphate Compounds

Name	See
Bromophos	Organophosphate Compounds
Bromophos-Ethyl	Organophosphate Compounds
Brush-Rhap	Chlorophenoxy Compounds
Bueno 6	Arsenicals
Bufencarb	Carbamates
Butylate	Dithiocarbamates and Thiocarbamates
Bux	Carbamates
CAMA	Arsenicals
Cacodylic Acid	Arsenicals
Calcium Acid Methanearsonate	Arsenicals
Calcium Arsenate	Arsenicals
Calcium Arsenite	Arsenicals
Caldon	Nitrophenolic and Nitrocresolic Herbicides
Carbamult	Carbamates
Carbaryl	Carbamates
Carbazinc	Dithiocarbamates and Thiocarbamates
Carbofuran	Carbamates
Carbophenothion	Organophosphate Compounds
Carzol	Carbamates
Celathion	Organophosphate Compounds
Certified Kiltrol-74 Weed Killer	Pentachlorophenol
Chem Bam	Dithiocarbamates and Thiocarbamates
Chem Pels C	Arsenicals
Chem-Sen 56	Arsenicals
Chemonite	Arsenicals
Chemox General	Nitrophenolic and Nitrocresolic Herbicides
Chemox PE	Nitrophenolic and Nitrocresolic Herbicides
Chemsect DNBP	Nitrophenolic and Nitrocresolic Herbicides
Chemsect DNOC	Nitrophenolic and Nitrocresolic Herbicides
Chipco Crab Kleen	Arsenicals
Chipco Thiram 75	Dithiocarbamates and Thiocarbamates
Chipco Turf Herbicide "D"	Chlorophenoxy Compounds
Chlordan	Organochlorines
Chlordane	Organochlorines
Chlorfenac	Chlorophenoxy Compounds
Chlorfenvinphos	Organophosphate Compounds
Chlormephos	Organophosphate Compounds
Chlorobenzilate	Organochlorines
Chlorophenothane	Organochlorines
Chloroxone	Chlorophenoxy Compounds
Chlorpyrifos	Organophosphate Compounds

(continued)

Name	See
Chlorthiophos	Organophosphate Compounds
Chrysron	Pyrethrins
Ciba-Geigy Ontrack (OS3, 4 or 5)	Pentachlorophenol
Cinerins	Pyrethrins
Ciodrin	Organophosphate Compounds
Co-Ral	Organophosphate Compounds
Copper Acetoarsenite	Arsenicals
Copper Arsenite	Arsenicals
Corozate	Dithiocarbamates and Thiocarbamates
Coumaphos	Organophosphate Compounds
Counter	Organophosphate Compounds
Crab-E-Rad	Arsenicals
Crisquat	Paraquat and Diquat
Crop Rider	Chlorophenoxy Compounds
Crotothane	Nitrophenolic and Nitrocresolic Herbicides
Crotoxyphos	Organophosphate Compounds
Cuman	Dithiocarbamates and Thiocarbamates
Curacron	Organophosphate Compounds
Curamil	Organophosphate Compounds
Cyanofenphos	Organophosphate Compounds
Cyanophos	Organophosphate Compounds
Cyanox	Organophosphate Compounds
Cycloate	Dithiocarbamates and Thiocarbamates
Cygon	Organophosphate Compounds
Cyolane	Organophosphate Compounds
Cypermethrin	Pyrethrins
Cythioate	Organophosphate Compounds
Cythion	Organophosphate Compounds
D 50	Chlorophenoxy Compounds
D(2, 4-D)	Chlorophenoxy Compounds
DB(2, 4-DB)	Chlorophenoxy Compounds
DDT	Organochlorines
DDVP	Organophosphate Compounds
DEF	Organophosphate Compounds
DMA	Arsenicals
DMA 100	Arsenicals
DMA 4	Chlorophenoxy Compounds
DNAP	Nitrophenolic and Nitrocresolic Herbicides
DNBP	Nitrophenolic and Nitrocresolic Herbicides
DNC	Nitrophenolic and Nitrocresolic Herbicides
DNOC	Nitrophenolic and Nitrocresolic Herbicides
DP-2 Antimicrobial	Pentachlorophenol

Name	See
DSE	Dithiocarbamates and Thiocarbamates
DSMA	Arsenicals
Dacamine 2D/2T	Chlorophenoxy Compounds
Dacamine 4D	Chlorophenoxy Compounds
Dacamine 4T	Chlorophenoxy Compounds
Daconate 6	Arsenicals
Dal-E-Rad	Arsenicals
Dalf	Organophosphate Compounds
Dapacryl	Nitrophenolic and Nitrocresolic Herbicides
Dasanit	Organophosphate Compounds
De Broussaillant 600	Chlorophenoxy Compounds
De Broussaillant Concentre	Chlorophenoxy Compounds
De-Fend	Organophosphate Compounds
De-Green	Organophosphate Compounds
Decamethrin	Pyrethrins
Ded-Weed	Chlorophenoxy Compounds
Ded-Weed Brush Killer	Chlorophenoxy Compounds
Delnav	Organophosphate Compounds
Demeton	Organophosphate Compounds
Demeton-Methyl	Organophosphate Compounds
Desiccant L-10	Arsenicals
Desormone	Chlorophenoxy Compounds
Dessin	Nitrophenolic and Nitrocresolic Herbicides
Dextrone	Paraquat and Diquat
Dextrone X	Paraquat and Diquat
Dexuron	Paraquat and Diquat
Di-Syston	Organophosphate Compounds
Di-Tac	Arsenicals
Dialifor	Organophosphate Compounds
Diallate	Dithiocarbamates and Thiocarbamates
Diazinon	Organophosphate Compounds
Dibrom	Organophosphate Compounds
Dicamba	Chlorophenoxy Compounds
Dicarzol	Carbamates
Dichlofenthion	Organophosphate Compounds
Dichloroanisic Acid	Chlorophenoxy Compounds
Dichlorophenoxyacetic Acid (2, 4-)	Chlorophenoxy Compounds
Dichlorprop	Chlorophenoxy Compounds
Dichlorvos	Organophosphate Compounds
Dicofol	Organochlorines
Dicotox	Chlorophenoxy Compounds

(continued)

Name	See
Dicrotophos	Organophosphate Compounds
Dieldrin	Organochlorines
Dieldrite	Organochlorines
Dienochlor	Organochlorines
Dilic	Arsenicals
Dimecron	Organophosphate Compounds
Dimethoate	Organophosphate Compounds
Dimetilan	Carbamates
Dinitro	Nitrophenolic and Nitrocresolic Herbicides
Dinitro General	Nitrophenolic and Nitrocresolic Herbicides
Dinitro-3	Nitrophenolic and Nitrocresolic Herbicides
Dinitrocresol	Nitrophenolic and Nitrocresolic Herbicides
Dintrophenol	Nitrophenolic and Nitrocresolic Herbicides
Dinobuton	Nitrophenolic and Nitrocresolic Herbicides
Dinocap	Nitrophenolic and Nitrocresolic Herbicides
Dinofen	Nitrophenolic and Nitrocresolic Herbicides
Dinopenton	Nitrophenolic and Nitrocresolic Herbicides
Dinoprop	Nitrophenolic and Nitrocresolic Herbicides
Dinosam	Nitrophenolic and Nitrocresolic Herbicides
Dinoseb	Nitrophenolic and Nitrocresolic Herbicides
Dinosulfon	Nitrophenolic and Nitrocresolic Herbicides
Dinoterb	Nitrophenolic and Nitrocresolic Herbicides
Dinoterbon	Nitrophenolic and Nitrocresolic Herbicides
Dinoxol	Chlorophenoxy Compounds
Dioxathion	Organophosphate Compounds
Dipher	Dithiocarbamates and Thiocarbamates
Dipterex	Organophosphate Compounds
Diquat	Paraquat and Diquat
Disodium Methyl Arsonate	Arsenicals
Disulfoton	Organophosphate Compounds
Dithane (D14, M22, M45, or Z78)	Dithiocarbamates and Thiocarbamates
Dithione	Organophosphate Compounds
Dormone	Chlorophenoxy Compounds
Dotan	Organophosphate Compounds
Dow General Weed Killer	Nitrophenolic and Nitrocresolic Herbicides
Dow Pentachlorophenol	Pentachlorophenol
Dow Selective Weed Killer	Nitrophenolic and Nitrocresolic Herbicides
Dowicide EC-7	Pentachlorophenol
Drawinol	Nitrophenolic and Nitrocresolic Herbicides
Draza	Carbamates
Drinox	Organochlorines
Drupina 90	Dithiocarbamates and Thiocarbamates

Name	See
Dursban	Organophosphate Compounds
Dyanap	Nitrophenolic and Nitrocresolic Herbicides
Dyfonate	Organophosphate Compounds
Dylox	Organophosphate Compounds
Dynamyte	Nitrophenolic and Nitrocresolic Herbicides
E-Z-Off D	Organophosphate Compounds
EPBP	Organophosphate Compounds
EPN	Organophosphate Compounds
EPTC	Dithiocarbamates and Thiocarbamates
Ectiban	Pyrethrins
Ekamet	Organophosphate Compounds
Ekatin	Organophosphate Compounds
Elgetol 30	Nitrophenolic and Nitrocresolic Herbicides
Elgetol 318	Nitrophenolic and Nitrocresolic Herbicides
Emerald Green	Arsenicals
Emulsamine BK	Chlorophenoxy Compounds
Emulsamine E-3	Chlorophenoxy Compounds
Endosan	Nitrophenolic and Nitrocresolic Herbicides
Endosulfan	Organochlorines
Endrin	Organochlorines
Entex	Organophosphate Compounds
Envert DT, T or 171	Chlorophenoxy Compounds
Eptam	Dithiocarbamates and Thiocarbamates
Esgram	Paraquat and Diquat
Esteron 245	Chlorophenoxy Compounds
Esteron 99 Concentrate	Chlorophenoxy Compounds
Esteron Brush Killer	Chlorophenoxy Compounds
Esteron Four	Chlorophenoxy Compounds
Estone	Chlorophenoxy Compounds
Ethion	Organophosphate Compounds
Ethoprop	Organophosphate Compounds
Ethyl Parathion	Organophosphate Compounds
Ethylan	Organochlorines
Etrimfos	Organophosphate Compounds
Fac	Organophosphate Compounds
Famfos	Organophosphate Compounds
Famphur	Organophosphate Compounds
Far-Go	Dithiocarbamates and Thiocarbamates
Fenac	Chlorophenoxy Compounds
Fenamiphos	Organophosphate Compounds
Fence Rider	Chlorophenoxy Compounds

(continued)

Name	See
Fenitrothion	Organophosphate Compounds
Fenophosphon	Organophosphate Compounds
Fenothrin	Pyrethrins
Fenpropanate	Pyrethrins
Fensulfothion	Organophosphate Compounds
Fenthion	Organophosphate Compounds
Fenvalerate	Pyrethrins
Ferbam	Dithiocarbamates and Thiocarbamates
Ferberk	Dithiocarbamates and Thiocarbamates
Fermide 850	Dithiocarbamates and Thiocarbamates
Fernasan	Dithiocarbamates and Thiocarbamates
Fernesta	Chlorophenoxy Compounds
Fernimine	Chlorophenoxy Compounds
Fernoxone	Chlorophenoxy Compounds
Ferxone	Chlorophenoxy Compounds
Field Clean Weed Killer	Chlorophenoxy Compounds
Folex	Organophosphate Compounds
Fonofos	Organophosphate Compounds
Formetanate HCI	Carbamates
Formothion	Organophosphate Compounds
Formula 40	Chlorophenoxy Compounds
Forron	Chlorophenoxy Compounds
French Green	Arsenicals
Fuclasin Ultra	Dithiocarbamates and Thiocarbamates
Fuklasin	Dithiocarbamates and Thiocarbamates
Fungostop	Dithiocarbamates and Thiocarbamates
Furadan	Carbamates
G 28029	Organophosphate Compounds
Gammexane	Organochlorines
Gardon Termi Tox	Pentachlorophenol
Gardona	Organophosphate Compounds
Gebutox	Nitrophenolic and Nitrocresolic Herbicides
Gramonol	Paraquat and Diquat
Griffin Manex	Dithiocarbamates and Thiocarbamates
Guthion	Organophosphate Compounds
Gypsine	Arsenicals
HCH	Organochlorines
Hedonal	Chlorophenoxy Compounds
Heptachlor	Organochlorines
Heptenophos	Organophosphate Compounds
Herb-All	Arsenicals
Herbidal	Chlorophenoxy Compounds

Name	See
Hexadrin	Organochlorines
Hexaferb	Dithiocarbamates and Thiocarbamates
Hexathane	Dithiocarbamates and Thiocarbamates
Hexathir	Dithiocarbamates and Thiocarbamates
Hexazir	Dithiocarbamates and Thiocarbamates
Hi-Yield Desiccant H-10	Arsenicals
Hostaquick	Organophosphate Compounds
Hostathion	Organophosphate Compounds
Imidan	Organophosphate Compounds
Inverton 245	Chlorophenoxy Compounds
Iodofenphos	Organophosphate Compounds
Isofenphos	Organophosphate Compounds
Isoxathion	Organophosphate Compounds
Jasmolins	Pyrethrins
Jones Ant Killer	Arsenicals
Karathane	Nitrophenolic and Nitrocresolic Herbicides
Karphos	Organophosphate Compounds
Kelthane	Organochlorines
Kepone	Organochlorines
Kill-All	Arsenicals
Kiloseb	Nitrophenolic and Nitrocresolic Herbicides
Klean Krop	Nitrophenolic and Nitrocresolic Herbicides
Knockmate	Dithiocarbamates and Thiocarbamates
Kuron	Chlorophenoxy Compounds
Kwell	Organochlorines
Kypman 80	Dithiocarbamates and Thiocarbamates
Kypzin	Dithiocarbamates and Thiocarbamates
Lannate	Carbamates
Lawn-Keep	Chlorophenoxy Compounds
Lead Arsenate	Arsenicals
Leptophos	Organophosphate Compounds
Lindane	Organochlorines
Line Rider	Chlorophenoxy Compounds
Lithate	Chlorophenoxy Compounds
Lonacol	Dithiocarbamates and Thiocarbamates
Lorsban	Organophosphate Compounds
Lysoff	Organophosphate Compounds
MAA	Arsenicals
MAMA	Arsenicals
MC2188	Organophosphate Compounds
MCPA	Chlorophenoxy Compounds

(continued)

Name	See
MCPB	Chlorophenoxy Compounds
MCPB-Ethyl	Chlorophenoxy Compounds
MCPCA	Chlorophenoxy Compounds
MCPP	Chlorophenoxy Compounds
MSMA	Arsenicals
Macondray	Chlorophenoxy Compounds
Malathion	Organophosphate Compounds
Mancozeb	Dithiocarbamates and Thiocarbamates
Maneb	Dithiocarbamates and Thiocarbamates
Manebgan	Dithiocarbamates and Thiocarbamates
Manesan	Dithiocarbamates and Thiocarbamates
Manzate	Dithiocarbamates and Thiocarbamates
Manzeb	Dithiocarbamates and Thiocarbamates
Manzin	Dithiocarbamates and Thiocarbamates
Marlate	Organochlorines
Matacil	Carbamates
Mecoprop	Chlorophenoxy Compounds
Mercuram	Dithiocarbamates and Thiocarbamates
Merger 823	Arsenicals
Merphos	Organophosphate Compounds
Mesamate	Arsenicals
Mesurol	Carbamates
Metasystox	Organophosphate Compounds
Methamidophos	Organophosphate Compounds
Methane Arsonate	Arsenicals
Methane Arsonic Acid	Arsenicals
Methar	Arsenicals
Methidathion	Organophosphate Compounds
Methiocarb	Carbamates
Methomyl	Carbamates
Methoxychlor	Organochlorines
Methyl Parathion	Organophosphate Compounds
Mevinphos	Organophosphate Compounds
Mezene	Dithiocarbamates and Thiocarbamates
Miracle	Chlorophenoxy Compounds
Mirex	Organochlorines
Mitis Green	Arsenicals
Mobilawn	Organophosphate Compounds
Mocap	Organophosphate Compounds
Mofisal	Paraquat and Diquat
Monitor	Organophosphate Compounds
Mono-Calcium Arsenite	Arsenicals

Name	See
Monoammonium Methane Arsonate	Arsenicals
Monoammonium Methyl Arsonate	Arsenicals
Monocrotophos	Organophosphate Compounds
Monosodium Methyl Arsonate	Arsenicals
Morocide	Nitrophenolic and Nitrocresolic Herbicides
Nabam	Dithiocarbamates and Thiocarbamates
Nabasan	Dithiocarbamates and Thiocarbamates
Naled	Organophosphate Compounds
Namate	Arsenicals
Naptalam	Nitrophenolic and Nitrocresolic Herbicides
Naptro	Nitrophenolic and Nitrocresolic Herbicides
Neguvon	Organophosphate Compounds
Nemacur	Organophosphate Compounds
Neo-Pynamin	Pyrethrins
Nespor	Dithiocarbamates and Thiocarbamates
Netagrone 600	Chlorophenoxy Compounds
Nexagan	Organophosphate Compounds
Nexion	Organophosphate Compounds
Nitrador	Nitrophenolic and Nitrocresolic Herbicides
Nitropone C	Nitrophenolic and Nitrocresolic Herbicides
Nomersan	Dithiocarbamates and Thiocarbamates
Nudrin	Carbamates
Nuvanol-N	Organophosphate Compounds
OMPA	Organophosphate Compounds
Oftanol	Organophosphate Compounds
Ofunack	Organophosphate Compounds
Orthene	Organophosphate Compounds
Ortho Paraquat CL	Paraquat and Diquat
Ortho Triox Liquid Vegetation Killer	Pentachlorophenol
Oxamyl	Carbamates
PCP	Pentachlorophenol
Paraquat	Paraquat and Diquat
Paraquat Dichloride	Paraquat and Diquat
Parathion	Organophosphate Compounds
Paris Green	Arsenicals
Parzate	Dithiocarbamates and Thiocarbamates
Pathclear	Paraqaut and Diquat
Pebulate	Dithiocarbamates and Thiocarbamates
Pencal	Arsenicals
Penchlorol	Pentachlorophenol

(*continued*)

Name	See
Penite	Arsenicals
Pennamine D	Organophosphate Compounds
Penncap-M	Chlorophenoxy Compounds
Penta	Pentachlorophenol
Pentac	Organochlorines
Pentachlorophenol	Pentachlorophenol
Pentacon	Pentachlorophenol
Penwar	Pentachlorophenol
Permethrin	Pyrethrins
Perthan	Organochlorines
Phencapton	Organophosphate Compounds
Phenthoate	Organophosphate Compounds
Phorate	Organophosphate Compounds
Phosalone	Organophosphate Compounds
Phosdrin	Organophosphate Compounds
Phosfolan	Organophosphate Compounds
Phosmet	Organophosphate Compounds
Phosphamidon	Organophosphate Compounds
Phosvel	Organophosphate Compounds
Phoxim	Organophosphate Compounds
Phthalthrin	Pyrethrins
Phytar 560	Arsenicals
Piperonyl Butoxide	Pyrethrins
Pirimicarb	Carbamates
Pirimiphos-Methyl	Organophosphate Compounds
Pirimor	Carbamates
Planotox	Chlorophenoxy Compounds
Plantgard	Chlorophenoxy Compounds
Polyram M	Dithiocarbamates and Thiocarbamates
Polyram Ultra	Dithiocarbamates and Thiocarbamates
Polyram Z	Dithiocarbamates and Thiocarbamates
Pomarsol Forte	Dithiocarbamates and Thiocarbamates
Pomarsol Z Forte	Dithiocarbamates and Thiocarbamates
Pounce	Pyrethrins
Preeglone	Paraquat and Diquat
Premerge 3	Nitrophenolic and Nitrocresolic Herbicides
Priglone	Paraquat and Diquat
Priltox	Pentachlorophenol
Proban	Organophosphate Compounds
Prodalumnol Double	Arsenicals
Prodaram	Dithiocarbamates and Thiocarbamates
Profenofos	Organophosphate Compounds

Name	See
Prolate	Organophosphate Compounds
Promecarb	Carbamates
Propetamphos	Organophosphate Compounds
Propoxur	Carbamates
Propylthiopyrophosphate	Organophosphate Compounds
Prothoate	Organophosphate Compounds
Purina Insect Oil Concentrate	Pentachlorophenol
Pydrin	Pyrethrins
Pynamin	Pyrethrins
Pyrazophos	Organophosphate Compounds
Pyrethrum	Pyrethrins
Pyridaphenthion	Organophosphate Compounds
Quinalphos	Organophosphate Compounds
Rabon	Organophosphate Compounds
Rad-E-Cate 25	Arsenicals
Rapid	Carbamates
Reglone	Paraquat and Diquat
Reglox	Paraquat and Diquat
Resmethrin	Pyrethrins
Rhodia	Chlorophenoxy Compounds
Rhodia Low Volatile Brush Killer No. 2	Chlorophenoxy Compounds
Ripcord	Pyrethrins
Ro-Neet	Dithiocarbamates and Thiocarbamates
S-Seven	Organophosphate Compounds
Safrotin	Organophosphate Compounds
Salvo	Chlorophenoxy Compounds
Schradan	Organophosphate Compounds
Schweinfurt Green	Arsenicals
Security	Arsenicals
Selinon	Nitrophenolic and Nitrocresolic Herbicides
Sevin	Carbamates
Silvex	Chlorophenoxy Compounds
Sipar	Paraquat and Diquat
Sinituho	Pentachlorophenol
Sinox General	Nitrophenolic and Nitrocresolic Herbicides
Snip Fly Bands	Carbamates
Sodar	Arsenicals
Sodium Arsenate	Arsenicals
Sodium Arsenite	Arsenicals
Sodium Pentachlorophenate	Pentachlorophenol

(continued)

Name	See
Soprabel	Arsenicals
Spectracide	Organophosphate Compounds
Spontox	Chlorophenoxy Compounds
Spotrete	Dithiocarbamates and Thiocarbamates
Spotton	Organophosphate Compounds
Spra-Cal	Arsenicals
Spring-Bak	Dithiocarbamates and Thiocarbamates
Spritz-Hormin/2,4-D	Chlorophenoxy Compounds
Spritz-Hormit/2,4-D	Chlorophenoxy Compounds
Strobane	Organochlorines
Strobane-T	Organochlorines
Subitex	Nitrophenolic and Nitrocresolic Herbicides
Sulfotepp	Organophosphate Compounds
Sulprofos	Organophosphate Compounds
Sumithion	Organophosphate Compounds
Super Crab-E-Rad-Calar	Arsenicals
Super D Weedone	Chlorophenoxy Compounds
Super Dal-E-Rad-Calar	Arsenicals
Superormone Concentre	Chlorophenoxy Compounds
Supracide	Organophosphate Compounds
Surecide	Organophosphate Compounds
Surpass	Dithiocarbamates and Thiocarbamates
Sutan	Dithiocarbamates and Thiocarbamates
Swat	Organophosphate Compounds
Synthrin	Pyrethrins
Systox	Organophosphate Compounds
T(2,4,5-T)	Chlorophenoxy Compounds
TB(2,4,5-TB)	Chlorophenoxy Compounds
TEPP	Organophosphate Compounds
TMTDS	Dithiocarbamates and Thiocarbamates
TP(2,4,5-TP)	Chlorophenoxy Compounds
Talan	Nitrophenolic and Nitrocresolic Herbicides
Talbot	Arsenicals
Target MSMA	Arsenicals
Temik	Carbamates
Tempephos	Organophosphate Compounds
Terbufos	Organophosphate Compounds
Terraklene	Paraquat and Diquat
Tersan	Dithiocarbamates and Thiocarbamates
Tersan 75	Dithiocarbamates and Thiocarbamates
Tersan LSR	Dithiocarbamates and Thiocarbamates
Tetrachlorvinphos	Organophosphate Compounds

Name	See
Tetraethyl Pyrophosphate	Organophosphate Compounds
Tetramethrin	Pyrethrins
Tetrapom	Dithiocarbamates and Thiocarbamates
Thimet	Organophosphate Compounds
Thiodan	Organochlorines
Thioknock	Dithiocarbamates and Thiocarbamates
Thiometon	Organophosphate Compounds
Thiophos	Organophosphate Compounds
Thiotex	Dithiocarbamates and Thiocarbamates
Thiram	Dithiocarbamates and Thiocarbamates
Thiramad	Dithiocarbamates and Thiocarbamates
Thirasan	Dithiocarbamates and Thiocarbamates
Thiuramin	Dithiocarbamates and Thiocarbamates
Thylate	Dithiocarbamates and Thiocarbamates
Tiezene	Dithiocarbamates and Thiocarbamates
Tiguvon	Organophosphate Compounds
Tillam	Dithiocarbamates and Thiocarbamates
Tirampa	Dithiocarbamates and Thiocarbamates
Torak	Organophosphate Compounds
Tormona	Chlorophenoxy Compounds
TotaCol	Paraquat and Diquat
Toxakil	Organochlorines
Toxaphene	Organochlorines
Trametan	Dithiocarbamates and Thiocarbamates
Trans-Vert	Arsenicals
Transamine	Chlorophenoxy Compounds
Triallate	Dithiocarbamates and Thiocarbamates
Triazophos	Organophosphate Compounds
Tributon	Chlorophenoxy Compounds
Tricarbamix Z	Dithiocarbamates and Thiocarbamates
Trichlorfon	Organophosphate Compounds
Trichloronate	Organophosphate Compounds
Trichlorophenoxyacetic Acid (2,4,5-)	Chlorophenoxy Compounds
Trichlorophylacetic Acid (2,3,6-)	Chlorophenoxy Compounds
Trifocide	Nitrophenolic and Nitrocresolic Herbicides
Trifungol	Dithiocarbamates and Thiocarbamates
Trimangol	Dithiocarbamates and Thiocarbamates
Trinoxol	Chlorophenoxy Compounds
Trioxone	Chlorophenoxy Compounds
Tripomol	Dithiocarbamates and Thiocarbamates

(continued)

Name	See
Triscabol	Dithiocarbamates and Thiocarbamates
Trithion	Organophosphate Compounds
Tritoftorol	Dithiocarbamates and Thiocarbamates
Tuads	Dithiocarbamates and Thiocarbamates
Tubothane	Dithiocarbamates and Thiocarbamates
U 46	Chlorophenoxy Compounds
U 46 Special	Chlorophenoxy Compounds
Unicrop DNBP	Nitrophenolic and Nitrocresolic Herbicides
Vancide FE-95	Dithiocarbamates and Thiocarbamates
Vancide MZ-96	Dithiocarbamates and Thiocarbamates
Vancide Maneb 80	Dithiocarbamates and Thiocarbamates
Vapona	Organophosphate Compounds
Veg-I-Kill	Pentachlorophenol
Veon 245	Chlorophenoxy Compounds
Vernam	Dithiocarbamates and Thiocarbamates
Vernolate	Dithiocarbamates and Thiocarbamates
Vertac Dinitro Weed Killer	Nitrophenolic and Nitrocresolic Herbicides
Verton 2D	Chlorophenoxy Compounds
Verton 2T	Chlorophenoxy Compounds
Visko-Rhap	Chlorophenoxy Compounds
Visko-Rhap LV 2D-2T	Chlorophenoxy Compounds
Vydate	Carbamates
Warbex	Organophosphate Compounds
Weed Tox	Chlorophenoxy Compounds
Weed-B-Gon	Chlorophenoxy Compounds
Weed-E-Rad	Arsenicals
Weed-E-Rad 360	Arsenicals
Weed-E-Rad DMA Powder	Arsenicals
Weed-Hoe	Arsenicals
Weed-Rhap	Chlorophenoxy Compounds
Weedar	Chlorophenoxy Compounds
Weedol	Paraquat and Diquat
Weedone	Chlorophenoxy Compounds
Weedtrine-D	Paraquat and Diquat
Weedtrol	Chlorophenoxy Compounds
White Arsenic	Arsenicals
Wood Preserver	Pentachlorophenol
Wood Tox 140	Pentachlorophenol
Z-C Spray	Dithiocarbamates and Thiocarbamates
Zebtox	Dithiocarbamates and Thiocarbamates
Zerlate	Dithiocarbamates and Thiocarbamates
Zincmate	Dithiocarbamates and Thiocarbamates

Name	See
Zineb	Dithiocarbamates and Thiocarbamates
Zinosan	Dithiocarbamates and Thiocarbamates
Ziram	Dithiocarbamates and Thiocarbamates
Ziramvis	Dithiocarbamates and Thiocarbamates
Zirasan 90	Dithiocarbamates and Thiocarbamates
Zirberk	Dithiocarbamates and Thiocarbamates
Zirex 90	Dithiocarbamates and Thiocarbamates
Ziride	Dithiocarbamates and Thiocarbamates
Zolone	Organophosphate Compounds

Appendix D
Chemicals in
CAS Number Order

CAS NUMBER	SYMPTOMS	INHALATION	INGESTION	SKIN	EYES	NAME
00050-00-0	1	1	2	1	1	Formaldehyde
00052-68-6	23	3	2	11	9	Dipterex
00055-63-0	26	4	7	6	3	Nitroglycerine
00056-23-5	9	3	5	6	5	Carbon Tetrachloride
00056-38-2	23	3	2	11	9	Parathion
00056-81-5	8	3	12	6	5	Glycerin
00057-14-7	31	5	5	2	5	Dimethylhydrazine (1,1-)
00057-55-6	8	3	12	6	5	Propylene Glycol
00057-74-9	31	5	5	2	5	Chlordane
00058-89-9	31	5	5	2	5	Lindane
00060-29-7	6	3	4	5	4	Ethyl Ether
00060-34-4	11	4	7	6	3	Monomethylhydrazine
00062-38-4	15	5	9	2	6	Phenylmercuric Acetate
00062-53-3	11	4	7	6	3	Aniline
00062-73-7	23	3	2	11	9	ODVP
00064-17-5	8	3	12	6	5	Ethyl Alcohol
00064-18-6	1	1	1	1	1	Formic Acid
00064-19-7	1	1	1	1	1	Acetic Acid
00067-56-1	8	3	12	6	5	Methyl Alcohol

(continued)

CAS NUMBER	SYMPTOMS	INHALATION	INGESTION	SKIN	EYES	NAME
00067-63-0	8	3	12	6	5	Isopropyl Alcohol
00067-64-1	8	3	12	6	5	Acetone
00067-66-3	9	3	5	6	5	Chloroform
00067-72-1	9	3	5	6	5	Hexachloroethane
00071-23-8	8	3	12	6	5	Propyl Alcohol
00071-36-3	8	3	12	6	5	Aliphatic Alcohols—Butyl
00071-41-0	8	3	12	6	5	Aliphatic Alcohols—Amyl
00071-43-2	7	3	4	6	5	Benzene
00071-55-6	9	3	5	6	5	Methylchloroform
00074-82-8	6	3	0	5	4	Methane
00074-83-9	17	4	0	6	5	Methyl Bromide
00074-84-0	6	3	0	5	4	Ethane
00074-85-1	6	3	0	5	4	Ethylene
00074-86-2	6	3	0	5	4	Acetylene
00074-87-3	6	3	0	5	4	Chloromethane
00074-87-3	6	3	0	5	4	Methyl Chloride
00074-89-5	19	1	3	4	2	Methylamine
00074-90-8	25	8	11	8	5	Hydrocyanic Acid
00074-93-1	4	1	0	1	1	Methyl Mercaptan
00074-96-4	9	3	5	6	5	Halothane
00074-97-5	9	3	5	6	5	Chlorobromomethane
00074-98-6	6	3	0	5	4	Propane
00075-00-3	9	3	5	5	4	Chloroethane
00075-01-4	14	6	10	7	7	Polyvinyl Chloride
00075-01-4	9	3	0	6	5	Vinyl Chloride
00075-02-5	6	3	0	5	4	Vinyl Fluoride
00075-04-7	19	1	3	4	2	Ethylamine
00075-05-8	25	8	11	8	5	Acetonitrile
00075-07-7	1	1	2	1	1	Acetaldehyde
00075-09-2	9	3	5	5	4	Methylene Chloride
00075-15-0	1	1	2	1	1	Carbon Disulfide
00075-20-7	5	7	3	4	2	Calcium Carbide
00075-21-8	17	4	2	6	5	Ethylene Oxide
00075-24-1	28	6	0	10	8	Trimethylaluminum
00075-25-2	9	3	5	6	5	Bromoform
00075-29-6	9	3	5	6	5	Chloropropane
00075-31-0	19	1	3	4	2	Isopropylamine

CAS NUMBER	SYMPTOMS	INHALATION	INGESTION	SKIN	EYES	NAME
00075-34-3	9	3	5	5	4	Dichloroethane
00075-35-4	9	3	5	5	4	Vinylidene Chloride
00075-37-6	6	3	0	5	4	Difluoroethanes
00075-38-7	6	3	0	5	4	Vinylidene Fluoride
00075-38-7	6	3	0	5	4	Difluoroethylene
00075-43-4	6	3	0	5	4	Dichlorofluoromethane
00075-43-4	6	3	0	5	4	Freon 21
00075-44-5	1	1	0	1	1	Phosgene
00075-45-6	6	3	0	5	4	Chlorodifluoromethane
00075-45-6	6	3	0	5	4	Freon 22
00075-46-7	6	3	0	5	4	Trifluoromethane
00075-50-3	19	1	3	4	2	Trimethylamine
00075-52-5	7	3	4	6	5	Nitromethane
00075-56-9	17	4	2	6	5	Propylene Oxide
00075-65-0	8	3	12	6	5	Butanol
00075-68-3	6	3	0	5	4	Chlorodifluoroethane
00075-68-3	6	3	0	5	4	Chlorofluoroethane
00075-68-3	6	3	0	5	4	Freon 142b
00075-69-4	6	3	0	5	4	Freon 11
00075-69-4	6	3	0	5	4	Trichlorofluoromethane
00075-71-8	6	3	0	5	4	Dichlorodifluoromethane
00075-71-8	6	3	0	5	4	Freon 12
00075-72-9	6	3	0	5	4	Chlorotrifluoromethane
00075-72-9	6	3	0	5	4	Freon 13
00075-73-0	6	3	0	5	4	Freon 14
00075-73-0	6	3	0	5	4	Tetrafluoromethane
00075-74-1	18	5	9	6	6	Tetramethyllead
00075-86-5	25	8	11	8	5	Acetone Cyanohydrin
00076-01-7	9	3	5	6	5	Pentachloroethane
00076-03-9	1	1	1	1	1	Trichloroacetic Acid
00076-06-2	17	4	2	6	5	Chloropicrin
00076-11-9	9	3	5	6	5	Tetrachlorodifluoroethane
00076-12-0	9	3	5	6	5	Freon 112
00076-13-1	9	3	5	6	5	Freon 113
00076-13-1	9	3	5	6	5	Trichlorotrifluoroethane
00076-14-2	9	3	5	6	5	Freon 114

(*continued*)

CAS NUMBER	SYMPTOMS	INHALATION	INGESTION	SKIN	EYES	NAME
00076-14-2	9	3	5	6	5	Dichlorotetrafluoroethane
00076-15-3	9	3	0	6	5	Chloropentafluoroethane
00076-15-3	9	3	5	6	5	Freon 115
00076-16-4	6	3	0	5	4	Freon 116
00076-16-4	6	3	0	5	4	Hexafluoroethane
00076-22-2	31	5	5	2	5	Camphor
00077-78-1	1	1	1	1	1	Dimethyl Sulfate
00078-00-2	18	5	9	6	6	Tetraethyllead
00078-82-0	25	8	11	8	5	Isobutyronitrile
00078-84-2	1	1	1	1	1	Isobutyraldehyde
00078-87-5	9	3	5	6	5	Dichloropropane
00078-93-3	8	3	12	6	5	Methyl Ethyl Ketone
00079-00-5	9	3	5	6	5	Trichloroethane
00079-01-6	9	3	5	6	5	Trichloroethylene
00079-06-1	17	4	2	6	5	Acrylamide
00079-11-8	1	1	1	1	1	Chloroacetic Acid
00079-20-9	8	3	12	6	5	Methyl Acetate
00079-21-0	1	1	1	1	1	Peracetic Acid
00079-22-1	1	1	2	1	1	Methyl Chloroformate
00079-34-5	9	3	5	6	5	Tetrachloroethane
00079-38-9	6	3	0	5	4	Chlorotrifluoroethylene
00080-62-6	17	4	2	6	5	Methyl Methacrylate Monomer
00084-74-2	17	4	2	6	5	Dibutyl Phthalate
00085-00-7	30	4	2	6	5	Diquat
00085-44-9	3	7	2	2	6	Phthalic Anhydride
00087-86-5	26	4	7	6	3	Pentachlorophenol
00087-86-5	26	4	7	6	3	Pentachlorophenate
00088-72-2	11	4	7	6	3	Nitrotoluene
00088-85-7	26	4	7	6	3	DNBP
00088-89-1	17	4	2	6	5	Picric Acid
00091-08-7	17	4	2	6	5	Toluene 2,6-di-Isocyanate
00091-20-3	11	4	7	6	3	Naphthalene
00091-58-7	29	2	10	6	6	Chloronaphthalenes
00091-59-8	11	4	7	6	3	Naphthylamines
00092-15-9	11	4	7	6	3	Anisidines (ortho)
00092-52-4	29	2	10	7	6	Diphenyl
00092-87-5	11	4	7	6	3	Benzidine

CAS NUMBER	SYMPTOMS	INHALATION	INGESTION	SKIN	EYES	NAME
00095-38-5	19	1	3	4	2	Aliphatic Amines
00095-50-1	9	3	5	6	5	Dichlorobenzene
00095-53-4	11	4	7	6	3	Toluidine
00095-68-1	11	4	7	6	3	Xylidine
00096-10-6	28	6	0	10	8	Diethylaluminum Chloride
00096-33-3	17	4	2	6	5	Methyl Acrylate
00097-93-8	28	6	0	10	8	Triethylaluminum
00098-00-0	8	3	12	6	5	Furfural Alcohol
00098-01-1	17	4	2	6	5	Furfural
00098-51-1	9	3	5	6	5	Butyltoluene
00098-82-8	7	3	4	6	5	Cumene
00098-95-3	11	4	7	6	3	Nitrobenzene
00099-35-4	11	4	7	6	3	Trinitrobenzene
00100-00-5	11	4	7	6	3	Nitrochlorobenzene (p-)
00100-01-6	11	4	7	6	3	Nitroanilines
00100-37-8	17	4	2	6	5	Diethylaminoethanol
00100-41-4	8	3	12	6	5	Ethylbenzene
00100-42-5	17	4	2	6	5	Styrene
00100-44-7	1	1	2	1	1	Benzyl Chloride
00100-63-0	11	4	7	6	3	Phenylhydrazine
00100-65-2	11	4	7	6	3	Phenylhydroxylamine
00100-99-2	28	6	0	10	8	Triisobutylaluminum
00101-68-8	17	4	2	6	5	Methylenebis (Phenyl Isocyanate)
00103-09-3	8	3	12	6	5	Ethylhexyl Acetate
00104-94-9	11	4	7	6	3	Anisidines (para)
00105-60-2	3	7	2	2	6	Caprolactam
00106-50-6	1	1	2	1	1	Phenylenediamine (p-)
00106-51-4	1	1	2	1	1	Quinone
00106-89-8	17	4	2	6	5	Epichlorohydrin
00106-90-1	17	4	2	6	5	Glycidyl Acrylate
00106-92-3	17	4	2	6	5	Allyl Glycidyl Ether
00106-97-8	6	3	0	5	4	Butane
00106-99-0	6	3	0	5	4	Butadiene
00107-02-8	1	1	2	1	1	Acrolein
00107-05-1	17	4	2	6	5	Allyl Chloride

(*continued*)

CAS NUMBER	SYMPTOMS	INHALATION	INGESTION	SKIN	EYES	NAME
00107-06-2	9	3	5	5	4	Ethylene Dichloride
00107-07-3	8	3	12	6	5	Ethylene Chlorohydrin
00107-10-8	19	1	3	4	2	Propylamine
00107-13-1	25	8	11	8	5	Acrylonitrile
00107-18-6	17	4	2	6	5	Allyl Alcohol
00107-20-0	1	1	1	1	1	Chloroacetaldehyde
00107-21-1	8	3	12	6	5	Ethylene Glycol
00107-27-7	15	5	9	2	6	Ethylmercuric Chloride
00107-49-3	23	3	2	11	9	TEPP
00107-98-2	17	4	2	6	5	Propylene Glycol Monomethyl Ether
00108-05-4	8	3	12	6	5	Vinyl Acetate
00108-10-1	8	3	12	6	5	Methyl Isobutyl Ketone
00108-21-4	8	3	12	6	5	Isopropyl Acetate
00108-24-7	1	1	1	1	1	Acetic Anhydride
00108-31-6	1	1	2	1	1	Maleic Anhydride
00108-46-3	1	1	1	1	1	Resorcinol
00108-82-7	8	3	12	6	5	Diisobutylcarbinol
00108-88-3	7	3	4	6	5	Toluene
00108-90-7	9	3	5	6	5	Chlorobenzene
00108-93-0	17	4	2	6	5	Cyclohexanol
00108-94-1	17	4	2	6	5	Cyclohexanone
00108-95-2	1	1	1	1	1	Phenol
00109-60-4	8	3	12	6	5	Propyl Acetate
00109-66-0	7	3	4	6	5	Pentane
00109-73-9	19	1	3	4	2	Butylamine
00109-77-3	25	8	11	8	5	Malononitrile
00109-86-4	8	3	12	6	5	Ethylene Glycol Monomethyl Ether
00109-89-7	19	1	3	4	2	Diethylamine
00110-12-3	8	3	12	6	5	Methyl Isoamyl Ketone
00110-19-0	8	3	12	6	5	Isobutyl Acetate
00110-54-3	7	3	4	6	5	Hexane
00110-82-7	7	3	4	6	5	Cyclohexane
00111-27-3	8	3	12	6	5	Hexanol
00111-30-8	17	4	2	6	5	Glutaraldehyde
00111-46-6	8	3	12	6	5	Diethylene Glycol

CAS NUMBER	SYMPTOMS	INHALATION	INGESTION	SKIN	EYES	NAME
00111-69-3	25	8	11	8	5	Adiponitrile
00111-84-2	7	3	4	6	5	Nonane
00111-86-4	7	3	4	6	5	Octane
00111-92-2	19	1	3	4	2	Dibutylamine
00112-27-6	8	3	12	6	5	Triethylene Glycol
00112-30-1	8	3	12	6	5	Decanol
00115-07-1	6	3	0	5	4	Propylene
00116-06-3	23	3	2	11	9	Aldicarb
00116-14-3	6	3	0	5	4	Tetrafluoroethylene
00118-52-5	1	1	2	1	1	Dichloro-5,5-Dimethylhydantoin
00118-74-1	29	2	10	6	6	Hexachlorobenzene
00118-96-7	11	4	7	6	3	Trinitrotoluene
00119-93-7	11	4	7	6	3	Tolidine (o-)
00121-14-2	11	4	7	6	3	Dinitrotoluene
00121-44-8	19	1	3	4	2	Triethylamine
00121-69-7	11	4	7	6	3	Dimethylaniline
00121-75-5	23	3	2	11	9	Malathion
00122-39-4	29	2	10	6	6	Diphenylamine
00123-31-9	11	4	7	6	3	Hydroquinone
00123-38-6	1	1	1	1	1	Propionaldehyde
00123-42-2	8	3	12	6	5	Diacetone Alcohol
00123-72-8	1	1	1	1	1	Butyraldehyde
00123-73-9	1	1	2	1	1	Crotonaldehyde
00123-86-4	8	3	12	6	5	Butyl Acetate
00123-91-1	8	3	12	6	5	Dioxane
00124-18-5	7	3	4	6	5	Decane
00124-38-9	6	3	0	5	4	Carbon Dioxide
00124-40-3	19	1	3	4	2	Dimethylamine
00126-73-8	1	1	2	1	1	Tributyl Phosphate
00126-98-7	25	8	11	8	5	Methacrylonitrile
00127-18-4	9	3	5	6	5	Tetrachloroethylene
00135-88-6	11	4	7	6	3	Phenylnaphthylamine
00140-88-5	17	4	2	6	5	Ethyl Acrylate
00141-43-5	19	1	3	4	2	Ethanolamine
00141-78-6	8	3	12	6	5	Ethyl Acetate

(continued)

CAS NUMBER	SYMPTOMS	INHALATION	INGESTION	SKIN	EYES	NAME
00142-82-5	7	3	4	6	5	Heptane
00142-84-7	19	1	3	4	2	Dipropylamine
00144-55-8	13	6	8	2	6	Sodium Bicarbonate
00144-62-7	1	2	1	1	1	Oxalic Acid
00151-50-8	25	8	11	8	5	Potassium Cyanide
00151-56-4	17	4	2	6	5	Ethyleneimine
00152-16-9	23	3	2	11	9	OMPA
00298-00-0	23	3	2	11	9	Methyl Parathion
00298-02-2	23	3	2	11	9	Phorate
00299-84-3	23	3	2	11	9	Ronnel
00302-01-2	31	4	5	6	5	Hydrazine
00309-00-2	31	5	5	2	5	Aldrin
00311-45-5	23	3	2	11	9	Paraoxon
00329-71-5	26	4	7	6	3	Dinitrophenols
00333-41-5	23	3	2	11	9	Diazinon
00334-88-3	17	4	2	6	5	Diazomethane
00353-36-6	6	3	0	5	4	Ethyl Fluoride
00371-86-8	23	3	2	11	9	Isopestox
00463-51-4	17	4	0	6	5	Ketene
00471-34-1	14	6	10	7	7	Calcium Carbonate
00479-45-8	17	4	2	6	5	Tetryl
00497-19-8	20	2	3	9	3	Sodium Carbonate
00500-28-7	23	3	2	11	9	Chlorthion
00504-29-0	31	5	5	2	5	Aminopyridine
00506-68-3	25	8	11	8	5	Cyanogen Bromide
00506-77-4	25	8	11	8	5	Cyanogen Chloride
00506-78-5	25	8	11	8	5	Cyanogen Iodide
00506-87-6	20	2	3	9	3	Ammonium Carbonate
00509-14-8	11	4	7	6	3	Tetranitromethane
00513-77-9	10	2	6	2	6	Barium Carbonate
00532-27-4	17	4	2	6	5	Chloroacetophenone (2-)
00534-52-1	26	4	7	6	3	DNOC
00540-59-0	9	3	5	6	5	Dichloroethylene
00540-72-7	27	6	0	8	5	Sodium Thiocyanate
00541-41-3	1	1	2	1	1	Ethyl Chloroformate
00542-88-1	17	4	2	6	5	Bis (Chloromethyl) Ether
00543-49-7	8	3	12	6	5	Heptanol

CAS NUMBER	SYMPTOMS	INHALATION	INGESTION	SKIN	EYES	NAME
00543-80-6	10	2	6	2	6	Barium Acetate
00552-30-7	3	7	2	2	6	Trimellitic Anhydride
00554-13-2	20	2	3	9	3	Lithium Carbonate
00554-84-7	26	4	7	6	3	Nitrophenols
00554-84-7	26	4	7	6	3	Nitrophenolics
00556-52-5	17	4	2	6	5	Glycidol
00557-98-2	9	3	5	6	5	Chloropropene
00562-95-8	18	5	9	6	6	Triethyllead
00563-80-4	8	3	12	6	5	Methyl Isopropyl Ketone
00564-00-1	17	4	2	6	5	Diepoxybutane
00584-08-7	20	2	3	9	3	Potassium Carbonate
00584-84-9	17	4	2	6	5	Toluene 2,4-di-Isocyanate
00591-78-6	8	3	12	6	5	Methyl n-Butyl Ketone
00593-53-3	6	3	0	5	4	Fluoromethane
00593-70-4	6	3	0	5	4	Chlorofluoromethane
00593-74-8	15	5	9	2	6	Dimethylmercury
00594-42-3	17	4	2	6	5	Perchloromethyl Mercaptan
00597-64-8	22	5	9	2	6	Tetraethyltin
00598-58-3	26	4	7	6	3	Methyl Nitrate
00598-63-0	24	6	2	2	6	Lead Carbonate
00600-25-9	17	4	2	6	5	Chloro-1-Nitropropane (1-)
00624-83-9	17	4	2	6	5	Methyl Isocyanate
00625-58-1	26	4	7	6	3	Ethyl Nitrate
00627-13-4	26	4	7	6	3	Propyl Nitrate
00627-44-1	15	5	9	2	6	Diethylmercury
00628-63-7	8	3	12	6	5	Amyl Acetate
00628-86-4	15	5	9	2	6	Mercury Fulminate
00628-96-6	26	4	7	6	3	Ethylene Glycol Dinitrate
00630-08-0	6	3	0	0	0	Carbon Monoxide
00631-60-7	16	5	2	2	6	Mercury Acetate
00761-44-4	22	5	9	2	6	Tripropyltin
00786-19-6	23	3	2	11	9	Trithion
00892-20-6	22	5	9	2	6	Triphenyltin
01120-46-3	18	5	9	6	6	Lead Oleate
01184-57-2	15	5	9	2	6	Methylmercury Hydroxide
01303-96-4	3	7	2	2	6	Sodium Borate

(continued)

CAS NUMBER	SYMPTOMS	INHALATION	INGESTION	SKIN	EYES	NAME
01304-28-5	10	2	6	2	6	Barium Oxide
01305-62-0	20	2	3	9	3	Calcium Hydroxide
01305-78-8	5	7	3	4	2	Calcium Oxide
01305-78-8	5	7	3	4	2	Lime
01309-45-1	2	1	1	3	1	Fluosilicic Acid
01309-60-0	24	6	2	2	6	Lead Dioxide
01309-60-0	24	6	2	2	6	Lead Oxide (PbO)
01310-58-3	5	7	3	4	2	Potassium Hydroxide
01310-73-2	5	7	3	4	2	Sodium Hydroxide
01313-59-3	5	7	3	4	2	Sodium Oxide
01313-60-6	5	7	3	4	2	Sodium Peroxide
01314-41-6	24	6	2	2	6	Lead Oxide (red)
01314-80-3	4	1	0	1	1	Phosphorus Pentasulfide
01314-87-0	24	6	2	2	6	Lead Sulfide
01319-77-3	1	1	2	1	1	Creosols
01330-20-7	7	3	4	6	5	Xylene
01332-17-8	3	7	2	2	6	Chlorinated Lime
01332-21-4	14	6	10	7	7	Asbestos
01332-58-7	14	6	10	7	7	Kaolin
01333-86-4	14	6	10	7	7	Carbon Black
01335-31-5	16	5	2	2	6	Mercury Oxycyanide
01335-32-6	24	6	2	2	6	Lead Subacetate
01335-85-9	26	4	7	6	3	Dinitrocresols
01336-21-6	19	1	3	4	2	Ammonium Hydroxide
01336-36-3	29	2	10	6	6	Polychlorinated Biphenyls (PCBs)
01344-28-1	14	6	10	7	7	Aluminum Oxide
01344-67-8	3	7	2	2	6	Copper Chloride
01461-25-2	22	5	9	2	6	Tetrabutyltin
01910-42-5	30	4	5	6	5	Paraquat
02104-64-5	23	3	2	11	9	EPN
02176-98-9	22	5	9	2	6	Tetrapropyltin
02179-59-1	17	4	2	6	5	Allyl Propyl Disulfide
02238-07-5	17	4	2	6	5	Diglycidyl Ether
02426-08-6	17	4	2	6	5	Butyl Glycidyl Ether (n-)
02587-84-0	18	5	9	6	6	Dibutyllead
02698-41-1	17	4	2	6	5	Chlorobenzylidene Malonitrile

CAS NUMBER	SYMPTOMS	INHALATION	INGESTION	SKIN	EYES	NAME
03333-52-6	25	8	11	8	5	Tetramethyl Succinonitrile
03689-24-5	23	3	2	11	9	Sulfotepp
03811-04-9	12	2	8	2	6	Potassium Chlorate
04685-14-7	30	4	2	6	5	Gramoxone
06032-29-7	8	3	12	6	5	Pentanol
06834-92-0	20	2	3	9	3	Sodium Silicate
07428-48-0	18	5	9	6	6	Lead Stearate
07429-90-5	14	6	10	7	7	Aluminum (dust)
07439-92-1	24	6	2	2	6	Lead (dust and fumes)
07439-97-6	16	5	2	2	6	Mercury (metal)
07439-97-6	16	5	2	2	6	Mercury (soluble salts)
07440-02-0	17	4	2	6	5	Nickel (fumes and dust)
07440-06-4	3	7	2	2	6	Platinum and Compounds
07440-09-7	5	7	3	4	2	Potassium
07440-23-5	5	7	3	4	2	Sodium
07440-32-6	14	6	10	7	7	Titanium (dust and fumes)
07440-36-0	21	1	2	1	1	Antimony and Compounds
07440-38-2	21	1	1	2	6	Arsenic
07440-39-3	10	2	6	2	6	Barium (soluble salts)
07440-43-9	3	7	2	2	6	Cadmium (dust and fumes) (metal)
07440-44-0	14	6	10	7	7	Carbon
07440-61-1	3	7	2	2	6	Uranium and Compounds
07440-62-2	3	7	2	2	6	Vanadium and Compounds
07442-13-9	18	5	9	6	6	Trimethyllead
07446-09-5	1	1	1	1	1	Sulfur Dioxide
07446-11-9	1	1	1	1	1	Sulfur Trioxide
07446-70-0	3	7	2	2	6	Aluminum Chloride
07446-70-0	3	7	2	2	6	Aluminum Trichloride
07477-40-7	13	6	8	2	6	Potassium Chloride
07487-88-9	27	6	0	7	6	Magnesium Sulfate
07487-94-7	16	5	2	2	6	Mercuric Chloride
07546-30-7	16	5	2	2	6	Mercurous Chloride
07550-45-0	3	7	2	2	6	Titanium Chlorides
07553-56-2	1	1	2	1	1	Iodine
07558-79-4	27	6	0	7	6	Disodium Phosphate

(continued)

CAS NUMBER	SYMPTOMS	INHALATION	INGESTION	SKIN	EYES	NAME
07580-67-8	20	2	3	9	3	Lithium Hydride
07601-54-9	20	2	3	9	3	Trisodium Phosphate
07601-89-0	12	2	8	2	6	Sodium Perchlorate
07601-90-3	1	1	1	1	1	Perchloric Acid
07616-94-6	1	1	0	3	1	Perchloryl Fluoride
07631-86-9	14	6	10	7	7	Silica
07646-85-7	3	7	2	2	6	Zinc Chloride
07647-01-0	1	1	1	1	1	Hydrochloric Acid
07647-01-0	1	1	1	1	1	Hydrogen Chloride
07647-14-5	13	6	8	2	6	Sodium Chloride
07664-38-2	1	1	1	1	1	Phosphoric Acid
07664-39-3	2	1	1	3	1	Hydrofluoric Acid
07664-41-7	19	1	3	4	2	Ammonia
07664-93-9	1	1	1	1	1	Sulfuric Acid
07681-49-4	2	1	1	3	1	Sodium Fluoride
07681-52-9	3	7	2	2	6	Sodium Hypochlorite
07697-37-2	1	1	1	1	1	Nitric Acid
07719-12-2	21	1	1	1	1	Phosphorus Trichloride
07722-84-1	1	1	1	1	1	Hydrogen Peroxide
07723-14-0	21	1	1	1	1	Phosphorus
07726-95-6	1	1	1	1	1	Bromine
07727-37-9	6	3	0	5	4	Nitrogen
07738-94-5	3	7	1	2	6	Chromic Acid
07758-19-2	1	1	1	1	1	Sodium Chlorite
07758-94-3	3	7	2	2	6	Iron Chloride
07758-97-6	24	6	2	2	6	Lead Chromate (yellow)
07758-98-7	3	7	2	2	6	Copper Sulfate
07767-82-6	27	6	0	7	6	Sodium Sulfate
07772-98-7	27	6	0	7	6	Sodium Thiosulfate
07774-29-0	16	5	2	2	6	Mercuric Iodide (red)
07775-09-9	12	2	8	2	6	Sodium Chlorate
07778-50-9	3	7	2	2	6	Potassium Dichromate
07778-54-3	3	7	1	2	6	Calcium Hypochlorite
07778-74-7	12	2	8	2	6	Potassium Perchlorate
07782-41-4	2	1	1	3	1	Fluorine
07782-50-5	1	1	1	1	1	Chlorine
07782-99-2	1	1	1	1	1	Sulfurous Acid